Lecture Notes in Computer Science 12010

Rachid Benmansour · Angelo Sifaleras ·
Nenad Mladenović (Eds.)

Variable Neighborhood Search

7th International Conference, ICVNS 2019
Rabat, Morocco, October 3–5, 2019
Revised Selected Papers

 Springer

Editors
Rachid Benmansour ⓘ
Institut National de Statistique et
d'Economie Appliquée (INSEA)
Rabat, Morocco

Angelo Sifaleras ⓘ
University of Macedonia
Thessaloniki, Greece

Nenad Mladenović ⓘ
Khalifa University
Abu Dhabi, United Arab Emirates

ISSN 0302-9743   ISSN 1611-3349   (electronic)
Lecture Notes in Computer Science
ISBN 978-3-030-44931-5   ISBN 978-3-030-44932-2   (eBook)
https://doi.org/10.1007/978-3-030-44932-2

LNCS Sublibrary: SL1 – Theoretical Computer Science and General Issues

This Springer imprint is published by the registered company Springer Nature Switzerland AG
The registered company address is: Gewerbestrasse 11, 6330 Cham, Switzerland

Preface

This volume edited by Rachid Benmansour, Angelo Sifaleras, and Nenad Mladenović contains peer-reviewed papers from the 7th International Conference on Variable Neighborhood Search (ICVNS 2019) held in Rabat, Morocco, during October 3–5, 2019. The conference follows previous successful meetings that were held in Sithonia, Halkidiki, Greece (2018); Ouro Preto, Brazil (2017); Malaga, Spain (2016); Djerba, Tunisia (2014); Herceg Novi, Montenegro (2012); and Puerto de La Cruz, Tenerife, Spain (2005). This edition was organized by Rachid Benmansour, from the Institut National de Statistique et d'Economie Appliquée (Morocco), who was the conference chair, Nenad Mladenović, from Khalifa University (UAE), who was the general chair, and Pierre Hansen, from GERAD and HEC Montréal (Canada), who was the honorary chair.

The main goal of ICVNS 2019 was to provide a stimulating environment in which researchers coming from various scientific fields could share and discuss their knowledge, expertise, and ideas related to the VNS Metaheuristic and its applications. The location of ICVNS 2019 in Rabat, Morocco, allowed to combine academic presentations and social networking.

The following three plenary speakers shared their current research directions with the ICVNS 2019 participants:

- Martine Labbé, from the Université Libre de Bruxelles, Belgium, "Bilevel optimisation and pricing problems"
- Eduardo G. Pardo, from the Universidad Politécnica de Madrid, Spain, "Parallelization of Variable Neighborhood Search: the present and future paradigm"
- Said Salhi, from Kent Business School, UK, "Hybridization and Deep Learning: case of VNS & LNS for a class of routing problems"

Around 40 participants took part in the ICVNS 2019 conference and a total of 28 submissions were accepted for oral presentation. A total of 13 full papers were accepted for publication in this LNCS volume after thorough peer reviewing by the members of the ICVNS 2019 Program Committee. These papers describe recent advances in methods and applications of VNS. The editors thank all the participants at the conference for their contributions and for their continuous effort to disseminate VNS, and are grateful to the reviewers for preparing excellent reports. The editors wish to acknowledge the Springer LNCS editorial staff for their support during the entire publication process.

Finally, we express our gratitude to the organizers and sponsors of the ICVNS 2019 meeting, especially:

- The National Institute of Statistics and Applied Economics (INSEA)
- The EURO Working Group on Metaheuristics (EWG EU/ME)
- The Research Laboratory in Information Systems, Intelligent Systems and Mathematical Modeling (SI2M)

Their support is greatly appreciated for making ICVNS 2019 a great scientific event.

February 2020

Rachid Benmansour
Angelo Sifaleras
Nenad Mladenović

Organization

Conference Organizers

Pierre Hansen (Honorary Chair)	GERAD and HEC Montréal, Canada
Nenad Mladenović (General Chair)	Khalifa University, UAE
Rachid Benmansour (Conference Chair)	INSEA, Morocco

Program Chairs

Rachid Benmansour	INSEA, Morocco
Angelo Sifaleras	University of Macedonia, Greece
Nenad Mladenović	Khalifa University, UAE

Program Committee

Abdelhakim AïtZaï	USTHB, Algeria
Ali Allahverdi	Kuwait University, Kuwait
Ada Alvarez	Universidad Autónoma de Nuevo León, Mexico
Cláudio Alves	University of Minho, Portugal
Abdelghani Bekrar	UPHF, France
Rachid Benmansour	INSEA, Morocco
Emilio Carrizosa	University of Seville, Spain
Gilles Caporossi	HEC Montréal, Canada
Vitor Nazário Coelho	UFF, Brazil
Tatjana Davidović	Mathematical Institute SANU, Serbia
Karl Doerner	University of Vienna, Austria
Abraham Duarte	Universidad Rey Juan Carlos, Spain
David Duvivier	UPHF, France
Michel Gendreau	Polytechnique Montréal, Canada
Saïd Hanafi	UPHF, France
Pierre Hansen	GERAD Montréal, Canada
Bassem Jarboui	Emirates College of Technology, UAE
Abdeslam Kadrani	INSEA, Morocco
Yuri Kochetov	Novosibirsk State University, Russia
Yannis Marinakis	Technical University of Crete, Greece
María Belén Melián-Batista	University La Laguna, Spain
Nenad Mladenović	Mathematical Institute SANU, Serbia
Patrick Siarry	Paris-Est Créteil University, France
Panos Pardalos	University of Florida, USA

Contents

A Reduced Variable Neighborhood Search Approach for Feature Selection in Cancer Classification

Angelos Pentelas, Angelo Sifaleras[(✉)] [ID], and Georgia Koloniari

Department of Applied Informatics, School of Information Sciences,
University of Macedonia, 156 Egnatia Street, 54636 Thessaloniki, Greece
apentelas@uom.edu.gr, {sifalera,gkoloniari}@uom.gr

Abstract. In this work we propose a Reduced Variable Neighborhood Search (RVNS) algorithm, to handle the gene selection problem in cancer classification. RVNS is utilized as the search method and gene subsets obtained are evaluated by three learning algorithms, namely support vector machine, k-nearest neighbors, and random forest. Experiments are conducted on five publicly available cancer related datasets, all characterized by a small sample size to dimensionality ratio. Since RVNS seeks gene subsets that yield accurate predictions for all three aforementioned classifiers, the obtained results can be considered more reliable. To the best of our knowledge, the proposed methodology is innovative due to the fact that, it combines the Recursive Feature Elimination (RFE) heuristic with a RVNS algorithm. Despite the large size of the problem instances, the suggested feature selection scheme converges within reasonably short time, when compared to similar methods. Results indicate high performance for RVNS that, is further improved when the RFE method is applied as a pre-processing step.

Keywords: Reduced Variable Neighborhood Search · Feature selection · Cancer classification

1 Introduction

Compelling technological advances, along with a well-established existent theoretical background, shaped the era of Big Data and Artificial Intelligence. These terms, usually intertwined, imprint the development of tools capable of collecting and storing complex data, as well as methods for mining knowledge from them. Industry and organisations tested and adopted such techniques in a sense that data-driven decisions and operations carry less bias and are, thus, more reliable.

However, the aforementioned trend results in datasets complicated enough that it takes great computational effort for machines to analyze and makes impossible for human experts to interpret, e.g., microarray datasets. In an attempt of achieving a fair trade-off between leveraging all the available information and interpreting an objective's results, Feature Selection (FS) emerged.

© Springer Nature Switzerland AG 2020
R. Benmansour et al. (Eds.): ICVNS 2019, LNCS 12010, pp. 1–16, 2020.
https://doi.org/10.1007/978-3-030-44932-2_1

On a high level, FS can be considered as a technique that, ideally, maintains only relevant information, i.e., features, of a dataset about the imminent analysis' scope and discards the rest as irrelevant. Since the FS problem has been proven to be NP-hard [19] and, in addition, in [21] it is implied that the choice of an effective FS method is dataset dependent, various FS techniques have been proposed in the literature. These can be arranged into four groups, namely *Filters*, *Wrappers*, *Embedded*, and *Ensemble*. However, following the recent research studies within the field, an observable shift towards hybridized FS schemes is apparent [1,3,5,8,18]. In the next three paragraphs, all methods are shortly described within a classification task context.

The *Filter* methods rely only on the intrinsic data characteristics, i.e., statistical metrics. Such techniques benefit from a low time complexity and limit the risk of model over-fitting since they do not take the learning algorithm's performance into consideration. The latter can be proven one of their most significant drawbacks, since the predictive ability of a model is a significant concern for domain experts.

Wrappers include techniques that continuously search into the feature space, select a feature subset, evaluate its quality by, usually, one classifier and repeat this process until some stopping criteria are met. The selection of a feature subset is typically driven by an intelligent mechanism (e.g., metaheuristics) and is not randomized. Despite being computationally more expensive than the filter methods, these techniques yield more accurate results and manageable sized solutions. Nevertheless, wrappers seem to undergo the risk of model over-fitting.

Trying to balance the pros and cons of the aforementioned FS classes, *Embedded* methods emerged. As stated in [6], such methods use the core of the classifier to establish criteria to rank features. Finally, *Ensemble* techniques, acting like ensemble of classifiers, combine methods described above on the assumption that combining the output of multiple experts is better than the output of any single expert [6]. Nonetheless, both of the aforesaid techniques come with deficiencies. In particular, *Embedded* methods are generally driven by heuristic approaches, thus leading to insufficient exploration of the solution space. *Ensemble* FS schemes, on the other hand, require higher computational time than any single FS technique they incorporate does. Moreover, the contribution of each FS scheme to the final feature subset is not obvious and necessitates examination.

The purpose of this work is to propose an efficient search mechanism for gene selection in cancer classification that limits the drawbacks of wrapper FS techniques, i.e., the risk of model over-fitting and the high computational cost, while it manages to obtain accurate results. To this end, we implement a *Reduced Variable Neighborhood Search* (RVNS) algorithm that searches the solution space in a systematic, yet computationally light, manner. Solutions provided by the RVNS are shared across Support Vector Machine (SVM), k-Nearest Neighbor (k-NN) and Random Forest (RF) classifiers for evaluation and their average accuracy, along with the solution's number of selected genes, are taken into consideration by an appropriate evaluation function. As a result, the final gene subsets obtained by our algorithm yield accurate predictions for more than one learning algorithms and findings can be further used with more reliability.

In the rationale that population-based meta-heuristics have been extensively studied within the FS field, we provide a single-point search meta-heuristic algorithm (i.e., RVNS) that, performs exceptionally in terms of accuracy, final gene subset size, and convergence time. By applying the embedded Support Vector Machine - Recursive Feature Elimination (RFE) technique as a pre-processing step that significantly reduces the feature space, we suggest the RFE-RVNS hybrid method. Both RFE-RVNS and RVNS were tested on five high-dimensional cancer-related datasets, frequently used in cognate research papers.

The structure of this work is as follows. In Sect. 2, we discuss similar approaches within the gene selection problem, focusing on recent research work and the methods they utilize. Next, we introduce our methodology in Sect. 3. Section 4 presents the results of our methods on five datasets and a comparison with related well-performing algorithms is quoted. Last comes a short summary of our findings, as well as thoughts for future work and improvements, in Sect. 5.

2 Related Work

Focusing on recently conducted studies, in [1] authors implemented two wrapper methods, namely a Genetic Algorithm (GA) and a Geometrical Particle Swarm Optimization (GPSO) to address the gene selection problem. The proposed FS schemes use SVM as their learning algorithm which obtains noteworthy results, after evaluating 4,000 solutions.

Another population-based approach is presented in the work of Alshamlan et al. [3]. A Genetic Bee Colony (GBC) optimization algorithm is applied on a reduced solution space, provided by the Maximum Relevance Minimum Redundancy (MRMR) filter method. SVM's accuracy is again selected as the primary optimization parameter. The overall performance of the hybridized technique is considered acceptable in terms of predictive capability and gene subset size. However, parameter values indicate the requirement of great computational effort, since more than 8,000 evaluations occur.

In a more recent study [5], two hybrid algorithms are presented combining both filter and wrapper FS methods. These two proposed approaches consist of a pre-selection phase, carried out by filter techniques, followed by a search phase that determines a good subset of genes for the classification. A wrapper metaheuristic is responsible for the latter. From an accuracy standpoint, results in eight datasets indicate competitive performance. The computational effort, though, proves underwhelming, with tens of minutes and even hours of runtime. Worth noticing, the classifiers utilized in the two methods are SVM and k-NN, respectively.

Finally, valuable insights come from [18], where authors combine the SVM-RFE embedded method with the MRMR filter one. The novelty of this research work is that, genes are ranked by a convex combination of the relevance given by SVM weights and the MRMR criterion. Results in this case are also acceptable, even though gene subset sizes can not be considered small enough.

With all referenced studies being after 2007, a trend towards hybridized FS schemes becomes apparent. More specifically we note that, filters and embedded methods are in many cases used as a pre-processing step in order to reduce the vast solution space of the gene subset selection problem. Afterwards, wrappers' advantages being exploited, producing small and informative gene subsets. Concerning the learning algorithms used, SVM and k-NN have been the most popular choices.

3   Research Methodology

In this section, we elaborate on all algorithms used within our research, as well as how they are combined to form the proposed RVNS and RFE-RVNS FS schemes.

3.1   Reduced Variable Neighborhood Search

Variable Neighborhood Search (VNS) is a metaheuristic method based on systematic changes in the neighborhood structure within a search, for the solution of various optimization problems. A large number of successful applications of VNS have already been proposed in the literature, [17,23]. In the years following, several variations of VNS emerged, with Reduced VNS (RVNS) being one of them. The essential difference between VNS and RVNS is that, the latter avoids any kind of local search within each neighborhood structure, as shown in Algorithm 1. This fact results in RVNS being computationally lighter than the basic algorithm and, thus; a promising search strategy in large problem instances.

Algorithm 1: RVNS pseudocode for a minimization problem

initialize solution x
while *stopping criteria are not met* **do**
$\quad$ $k = 1$
$\quad$ **while** $k \leq k_{max}$ **do**
$\quad\quad$ generate x' a random solution from neighborhood $N_k(x)$
$\quad\quad$ **if** *evaluate(x') $<$ evaluate(x)* **then**
$\quad\quad\quad$ $x = x'$
$\quad\quad\quad$ $k = 1$
$\quad\quad$ **else**
$\quad\quad\quad$ $k = k + 1$
$\quad\quad$ **end**
$\quad$ **end**
end
return x;

Each candidate solution s is represented as a binary, 1-dimensional array of length N, with N denoting the number of genes in each dataset. For instance, a candidate solution in a dataset with five genes could be: $s = [0, 1, 1, 0, 1]$ which means that, the second, third, and the fifth genes of the dataset are selected; while the first and the fourth are not.

Furthermore, the three following neighborhood structures (i.e., $k_{max} =$ three) are used by both RFE-RVNS and RVNS schemes:

1. *Replace a selected gene of the incumbent solution with an un-selected one.*
2. *Replace two selected genes of the incumbent solution with an un-selected one. If the incumbent solution has only one gene selected, return the incumbent solution.*
3. *Add an un-selected gene to the incumbent solution. If there are no more genes to add, return the incumbent solution.*

The neighborhood order, which is also decisive, is as indicated above. In this manner, RVNS first tries to improve the current solution by keeping the same number of selected genes and, in case that fails, moves to the second neighborhood that reduces the selected genes by one. It is only when both these strategies are unsuccessful that the algorithm will seek a new solution with more selected genes. It should be pointed out that, in all experiments, the initial solution is generated arbitrarily with two randomly selected genes. Therefore, according to the neighborhood definitions above, no exception-handling is required for the case of zero selected genes.

Example 3.1 Assume a microarray dataset with five genes and an incumbent solution $s = [0, 1, 1, 0, 1]$. Let us denote with $N_i(s)$, $i \in \{1, 2, 3\}$, the sets of neighboring solutions of s. According to the three neighborhood structures as defined above, three resulting solutions could be $s_1 = [0, 1, 1, 1, 0] \in N_1(s)$, $s_2 = [1, 0, 1, 0, 0] \in N_2(s)$ and $s_3 = [1, 1, 1, 0, 1] \in N_3(s)$.

3.2   Recursive Feature Elimination

Recursive Feature Elimination (RFE) is a heuristic feature ranking approach that determines the importance of each feature based on a learning model's *coefficient* attribute or a *feature importance* metric. RFE is capable of yielding subsets with a specified number of features by repeatedly removing the least significant one(s).

Appertaining to the embedded FS techniques, RFE needs to be associated with a learning algorithm in order to be meaningful. Authors in [14], who introduced the RFE algorithm, combined it with an SVM classifier and successfully tested their SVM-RFE method on two microarray datasets.

In Algorithm 2, the process that SVM-RFE follows to rank all features is given. More specifically, at each iteration, the least significant feature is removed from the *survivable features* vector (i.e., s) and is appended to the *ranked list of features* (i.e, r) one. The necessity of each feature is quantified by the extent of contribution it occupies in the learning model. In the case of SVM, the importance of each feature is calculated through the w and c vectors, as illustrated in the aforementioned algorithm.

Algorithm 2: SVM-RFE pseudocode

Input: $X_0 = [x_1, x_2, ..., x_k, ..., x_l]^T$ `// training examples`
Input: $y = [y_1, y_2, ..., y_k, ..., y_l]^T$ `// class labels`
initialize subset of surviving features $s = [1, 2, ..., n]$
initialize feature ranked list $r = [\,]$
while $s \neq \emptyset$ **do**

    $X = X_0(:, s)$ `// restrict training examples`
    $\alpha = $ SVM-train(X, y) `// train the classifier`
    $w = \sum_k \alpha_k y_k x_k$ `// compute the weight vector`
    $c_i = (w_i)^2, \forall i$ `// compute the ranking criteria`
    $f = argmin(c)$ `// find the feature with the smallest ranking`
    $r = [s(f), r]$ `// update feature ranked list`
    $s = s(1 : f - 1, f + 1 : length(s))$ `/* eliminate the feature with`
        `smallest ranking criterion`   `*/`

end
return r;

3.3 Learning Algorithms

Support Vector Machine. In [9], Cortes and Vapnik proposed a remarkably effective learning algorithm called Support Vector Machine (SVM). SVM, conceptually implemented on a very simple idea, seeks for the surface, i.e., hyperplane, that can optimally segregate two-class training data. Predictions are based on what side of the, already defined, hyper-plane future data are mapped into. Note that SVMs can also be extended for multi-class classification tasks. Its simplicity, flexibility, and satisfactory computational complexity render SVMs superior to many supervised learning algorithms. As a result, several FS methods suggested in the literature have adopted the aforementioned classifier as their primary evaluation metric [1,3,5,11,14,18].

k-Nearest Neighbors. k-Nearest Neighbors (k-NN) is another powerful supervised learning algorithm widely used within the FS process [5,8,22]. It is considered a lazy learning algorithm, i.e., it does not make any assumptions about the underlying data distribution. Given a distance metric and a future data point mapped into the feature space, the class label assigned to the latter depends on the class labels of its k less-distant records. Leveraging mathematical topology's attributes, computation of the k-nearest neighbors can be efficiently achieved.

Random Forest. A Decision Tree (DT) is a logical structure consisting of parent and children nodes. In a high level approach, a splitting criterion is applied on each parent node in an attempt to yield pure children nodes, i.e., nodes that contain data points of one class, only. The Random Forest (RF) classifier improves the predictive capability of a single DT by incorporating many DTs that are built upon a random subset of data features. The class prediction of a future instance is justified by the majority of the partial class predictions each DT makes. RF is, thus, an ensemble of classifiers and demonstrates high performance in many machine learning applications, e.g., [4, 10].

3.4 Hybrid RFE-RVNS Method

In an attempt to enhance RVNS's performance, we apply the RFE heuristic approach as a pre-processing step. In that way, a significant number of possibly redundant genes are eliminated and the resulting solution space is handed over RVNS to search into. Therefore, a new search strategy is formed that we refer to as RFE-RVNS. Figure 1 depicts the aforementioned process.

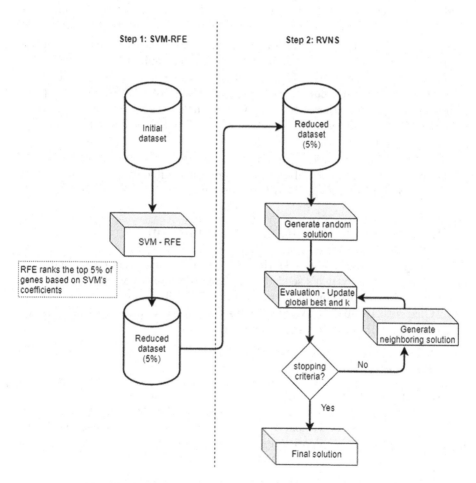

Fig. 1. The RFE-RVNS flowchart. The value of k indicates the neighborhood structure the algorithm is searching into.

Each candidate solution is evaluated by four metrics; the accuracy of the three classifiers and the size of the incumbent gene subset. Consequently, we define a fitness function that, is described below:

$$evaluate(s) = \alpha * \underbrace{\frac{3}{a_1(s) + a_2(s) + a_3(s)}}_{a(s)} + (1 - \alpha) * g(s) \qquad (1)$$

where $a_1(s)$, $a_2(s)$, and $a_3(s)$ denote the accuracy the SVM, k-NN and RF classifiers yield from solution s, respectively, and $g(s)$ indicates the number of selected genes. Assuming each predictive model performs at least as good as a random classification, $a_i(s) \in [0.5, 1.0], \forall i \in \{1, 2, 3\}$ (binary classification), thus $\frac{3}{\sum_{i=1,3} a_i(s)} \in [1, 2]$, while $g(s)$ is a positive integer restricted by the number of genes in each dataset. Parameter α acts as weight to the average accuracy of the three classifiers, while $1 - \alpha$ acts similarly to the gene subset size.

The evaluation function in Eq. 1 was selected since it can offer a good trade-off between the overall accuracy and the final number of selected genes. Experimentation led us to setting α to 0.99; a value that, is consistent both with our objective of finding informative gene subsets and with the co-domains of $\frac{3}{a(s)}$ and $g(s)$ in Eq. 1.

A similar fitness function is used in [1] and manages to balance accurate predictions and small gene subsets, although with different weight values and using the accuracy of a single learning algorithm.

4   Experimental Results and Comparison

All learning algorithms mentioned, the RFE heuristic, as well as the data normalization leverage the Python's scikit-learn library, developed for data science purposes. Experiments are conducted on an Intel i7-7700k 4-core processor, clocked at 4.2 Ghz, with 16 Gb of RAM. Single runs of both RVNS and RFE-RVNS never exceeded a minute, pre-processing included.

4.1   Parameter Settings

RVNS. Along with the neighborhood structures defined in Sect. 3, an essential parameter of RVNS that should be specified is the algorithm's termination criteria. In our implementation, we set those to be 300 iterations. The latter indicates that the RVNS algorithm evaluates exactly 300 solutions which is just as 900 classifications, i.e., three classifications per evaluation.

RFE. The RFE heuristic is applied with an SVM classifier. In each dataset, the SVM-RFE method eliminates 95% of the genes that, are considered irrelevant. The way of achieving this is by removing nine times a 10% (referring to the initial number of genes) of the least important genes and, finally, a 5%. Let us note that, in a typical RFE execution, such an elimination-step is considered quite large. In order to maintain the computational complexity low, and since RFE is not the primary search method, we apply it with the selected parameters we mentioned above.

Learning Algorithms. Core parameters of the learning algorithms are empirically selected with two ends in mind; accuracy performance and computational efficiency. Seeking for a balance between these two, we tested k-NN with k in $\{1, 3, 5, 7, 9\}$. Additionally, various RF implementations, with number of DT's in $\{10, 20, 30, 40, 50\}$ and pruning depth value in $\{10, 15, 20\}$, helped us proceed to our final choice.

The number of neighboring classes that k-NN takes into consideration before classifying an unknown patient is set to five and the RF classifier predicts class labels by consulting with 20 10-depth pruned decision trees. Moreover, the SVM classifier is implemented with a linear kernel meaning that, it searches for the best linear hyper-plane that is able to discriminate the data. The accuracy obtained from each learning algorithm is averaged after a 10-fold cross-validation.

4.2   Data Description and Preprocessing

The proposed methodology is tested on five publicly available cancer-related datasets; the Leukemia, Lung, Ovarian, Colon, and Breast cancer datasets. The first four were originally taken from the public Kent Ridge Bio-medical Data Repository, which is now hosted in the ELVIRA Biomedical Data Repository (http://leo.ugr.es/elvira/DBCRepository). The Breast Cancer Dataset was available under https://data.mendeley.com/datasets/v3cc2p38hb/1. Sample size, dimensionality, and the number of classes of each dataset are depicted in Table 1.

Table 1. Dataset characteristics

Dataset	Sample size	Number of genes	Number of classes	Reference
Leukemia	72	7,129	2	[12]
Lung	181	12,533	2	[13]
Ovarian	253	15,154	2	[20]
Colon	62	2,000	2	[2]
Breast	590	17,814	2	[7]

All data values are normalized and missing ones are replaced by zero's, i.e., their mean. It should be noted that, only in the Breast cancer dataset, a few missing values are found.

4.3   Performance of RFE-RVNS and RVNS

The performance of the proposed algorithms on the selected datasets is depicted in Tables 2, 3, 4, 5, and 6. The metrics measured are the *Best*, *Mean*, and *Worst* values of each of the classifiers' accuracy, along with the respective number of genes values (*#Genes*). The *Average accuracy* metric, which is measured as the average accuracy value of SVM, k-NN and RF in a single run, should not be interpreted as a typical learning algorithm's accuracy, but rather as the ability of the proposed algorithms to obtain informative genes for all classifiers.

Table 2. Performance of RFE - RVNS and RVNS algorithms when applied with the SVM, k-NN and RF classifiers on the Leukemia dataset after ten independent runs.

	RFE-RVNS			RVNS		
Metric	Best	Mean	Worst	Best	Mean	Worst
Average accuracy	99.58	98.88	97.64	97.44	94.26	89.35
SVM accuracy	100	98.75	94.58	100	95.18	86.67
k-NN accuracy	100	99.58	97.08	97.5	93.84	87.56
RF accuracy	100	98.32	97.08	97.5	93.76	87.2
#Genes	3	3.8	5	2	4.7	8

Table 3. Performance of RFE - RVNS and RVNS algorithms when applied with the SVM, k-NN and RF classifiers on the Lung cancer dataset after ten independent runs.

	RFE-RVNS			RVNS		
Metric	Best	Mean	Worst	Best	Mean	Worst
Average accuracy	99.44	98.65	97.22	98.55	95.67	91.88
SVM accuracy	99.44	98.73	97.22	98.36	95.75	91.17
k-NN accuracy	100	98.67	97.22	98.36	95.19	91.14
RF accuracy	100	98.56	97.22	98.92	96.07	93.33
#Genes	2	2.2	3	2	2.5	3

Commenting upon figures in Tables 2, 3, 4, and 6, RFE-RVNS managed to obtain a maximum, i.e., the maximum of bests, of 100% accuracy and a minimum, i.e., the minimum of worst, of 97.22%, while the corresponding values for RVNS are 99.10% and 89.35%, respectively. In the case of the Colon dataset, both RFE-RVNS and RVNS faced some adversities in finding small and informative gene subsets with a mean accuracy of 91.23% and 87.22%, respectively. Notable is the fact that, in the Ovarian dataset, RFE-RVNS managed to simultaneously yield 100% accuracy for all three classifiers.

Table 4. Performance of RFE - RVNS and RVNS algorithms when applied with the SVM, k-NN and RF classifiers on the Ovarian cancer dataset after ten independent runs.

Metric	RFE-RVNS			RVNS		
	Best	Mean	Worst	Best	Mean	Worst
Average accuracy	100	99.18	97.87	98.67	96.94	94.77
SVM accuracy	100	99.49	97.62	99.6	98.06	95.63
k-NN accuracy	100	99.21	98	98.4	97.04	94.52
RF accuracy	100	98.85	97.6	95.74	95.74	91.39
#Genes	2	2.4	3	3	4.1	7

Table 5. Performance of RFE - RVNS and RVNS algorithms when applied with the SVM, k-NN and RF classifiers on the Colon cancer dataset after ten independent runs.

Metric	RFE-RVNS			RVNS		
	Best	Mean	Worst	Best	Mean	Worst
Average accuracy	93.73	91.23	88.57	89.68	87.72	85.24
SVM accuracy	96.90	93.36	88.57	93.33	89.50	84.05
k-NN accuracy	96.90	92.69	89.05	90.00	87.79	84.05
RF accuracy	90.24	87.64	84.05	88.57	85.88	80.48
#Genes	4	5.5	8	3	5.5	8

Table 6. Performance of RFE - RVNS and RVNS algorithms when applied with the SVM, k-NN and RF classifiers on the Breast cancer dataset after ten independent runs.

Metric	RFE-RVNS			RVNS		
	Best	Mean	Worst	Best	Mean	Worst
Average accuracy	99.27	98.83	98.37	99.1	98.35	97.06
SVM accuracy	99.32	98.83	98.47	99.32	98.39	96.95
k-NN accuracy	99.32	98.97	98.31	99.32	98.39	97.12
RF accuracy	99.32	98.7	98.14	99.16	98.29	97.12
#Genes	1	1.7	2	2	2.5	3

Concerning the gene subset size, the mean number of selected genes is impressively small under both approaches, with 3.8-gene and 4.7-gene subsets being the largest average ones for RFE-RVNS and RVNS respectively. Again, in the Colon dataset, the behavior differs a little with slightly larger gene subsets.

While both methods perform worthy, not only in terms of yielding informative, to all classifiers, gene subsets, but also small sized ones, the dominance of RFE-RVNS over RVNS cannot be overlooked.

Questioning whether one learning algorithm is favored over the others, or whether their predictive ability significantly varies, Fig. 2 shows that only in the case of the Colon cancer dataset, the RF model performs somewhat worst.

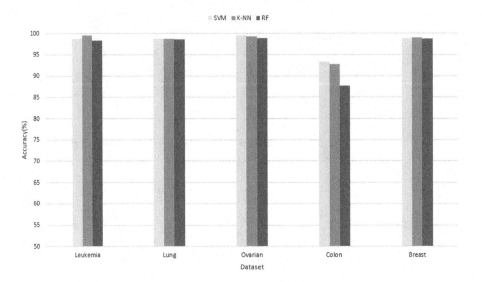

Fig. 2. SVM, k-NN and RF mean accuracy in each dataset obtained by RFE-RVNS.

Trying to decipher our method's behavior on the Colon dataset, we depict the classifiers' accuracy on current best solutions found every 15 iterations of a typical RVNS execution. The graphs illustrated in Fig. 3 indicate that, gene subsets obtained often improve one learning algorithm's performance but worsen another's, e.g., iterations 40 and 145. This phenomenon adds to our intention of implementing a gene selection strategy that returns informative gene subsets for more than one classifier, in the sense of quality and reliability.

4.4 Comparison

As stated earlier, the proposed fitness function tries to achieve high performance on three learning algorithms while maintaining a small gene subset size. However, most related work was conducted by targeting one or two ends. Thereby, within the context of a *search strategy* comparison, the objective function of RFE-RVNS and RVNS is modified in order to take only the SVM's accuracy into consideration, meaning that only a third of the classifications originally made will occur. That allows us to intensify the search capability of RFE-RVNS and RVNS by increasing the iterations from 300 to 500 and 1,000 respectively. Comparatively, in most related studies, wrapper methods tend to evaluate a few thousands candidate solutions as mentioned in Sect. 2.

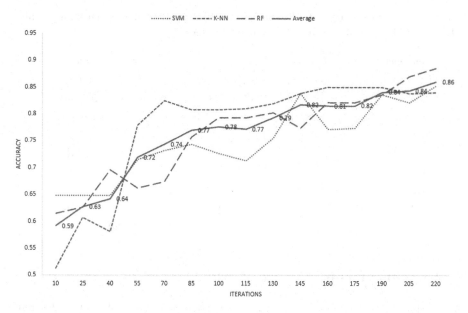

Fig. 3. SVM, k-NN, RF and average accuracy values from a typical execution of RVNS on the Colon dataset.

It is also in the same section that we refer to the Genetic Algorithm (GA) [1], the Geometrical Particle Swarm Optimization (GPSO) [1], the Genetic Bee Colony (GBC) [3] and the Maximum Relevance Minimum Redundancy - SVM-RFE (MRMR+SVM) [18] algorithms. Furthermore, in Table 7, performance of FS schemes like the Multiple Filter Multiple Wrapper (MFMW) [15] and the Ensemble Neural Network (ENN) [16] is presented. Lastly, it must be pointed out that, the Feature Selection - Random Projection (FS+RP) [24] method does not appertain to typical FS techniques presented in this paper; instead, it is associated with the feature extraction ones. However, we proceed to a comparison with it since, to the best of our knowledge, no other FS methods tested on the exact Breast cancer dataset can be found in the literature.

In Table 7, results indicate that RFE-RVNS outperforms well-known gene selection methods in all datasets except for the Colon one, while RVNS also obtains notable results. Thus, a small, yet informative, gene subset can be successfully obtained under a Variable Neighborhood Search strategy. Compared to similar methods, our algorithms require less amount of computational time since they evaluate significantly less candidate solutions.

Table 7. The performance of RFE-RVNS and RVNS algorithms, when applied only with the SVM classifier, compared to similar methods.

Reference	Leukemia	Lung	Ovarian	Breast	Colon
RFE-RVNS	99.86[4]	99.51[3]	99.80[3]	99.12[2]	96.69[5]
RVNS	98.84[5]	98.67[3]	98.55[4]	98.66[2]	93.74[6]
GA [1]	95.86[4]	99.49[4]	98.83[4]	-	100[3]
GPSO [1]	97.38[3]	99.00[4]	99.44[4]	-	100[2]
GBC [3]	96.43[5]	-	-	-	91.51[5]
MFMW [15]	-	98.34[6]	-	-	95.16[6]
MRMR+SVM [18]	98.35[37]	-	-	-	91.68[78]
ENN [16]	-	-	99.21[75]	-	81.48[-]
FS+RP [24]	-	-	-	98.97[>100]	-

5   Conclusions and Future Work

In this paper, our aim was to suggest an efficient wrapper feature selection method capable of yielding informative gene subsets for cancer classification. Therefore, we proposed a Reduced Variable Neighborhood Search algorithm as the primary search strategy. In many cases though, performance of different learning algorithms may significantly vary, despite learning from the same data (i.e., gene subsets). Consequently, we evaluated each gene subset by three classifiers, i.e., support vector machine, k-nearest neighbors and random forest, and balanced the extra computational effort by enforcing considerably less, compared to the literature, classification attempts. In addition to that, we applied the Recursive Feature Elimination heuristic method to reduce the feature space which was then given to RVNS to search into.

Both RFE-RVNS and RVNS performed well despite the large size of problem instances and the computationally intensive 3-model building. Results on five well-known publicly available microarray datasets indicate high performance of RVNS that manages to obtain high accuracy for all three classifiers while still keeping the gene subset size relatively small. By applying RFE and executing the RVNS algorithm on a significantly reduced feature space (5% of the initial size), the total performance is considerably improved. As a result, small-sized gene subsets obtained can be suggested to experts with higher reliability.

We conclude by acknowledging that, an algorithm's robustness constitutes an important performance criterion. The development of an appropriate initialization (construction) method might add to that direction. Further study on the latter, along with testing our method on more datasets and different domains (e.g., text classification) will concern us in future work.

Acknowledgement. The second author has been funded by the University of Macedonia Research Committee as part of the "Principal Research 2019" funding scheme (ID 81307).

References

1. Alba, E., Garcia-Nieto, J., Jourdan, L., Talbi, E.G.: Gene selection in cancer classification using PSO/SVM and GA/SVM hybrid algorithms. In: IEEE Congress on Evolutionary Computation (CEC 2007), pp. 284–290 (2007)
2. Alon, U., et al.: Broad patterns of gene expression revealed by clustering analysis of tumor and normal colon tissues probed by oligonucleotide arrays. Proc. Natl. Acad. Sci. **96**(12), 6745–6750 (1999)
3. Alshamlan, H.M., Badr, G.H., Alohali, Y.A.: Genetic bee colony (GBC) algorithm: a new gene selection method for microarray cancer classification. Comput. Biol. Chem. **56**, 49–60 (2015)
4. Belgiu, M., Drăguţ, L.: Random forest in remote sensing: a review of applications and future directions. ISPRS J. Photogrammetry Remote Sens. **114**, 24–31 (2016)
5. Bir-Jmel, A., Douiri, S.M., Elbernoussi, S.: Gene selection via BPSO and backward generation for cancer classification. RAIRO-Oper. Res. **53**(1), 269–288 (2019)
6. Bolón-Canedo, V., Sánchez-Marono, N., Alonso-Betanzos, A., Benítez, J.M., Herrera, F.: A review of microarray datasets and applied feature selection methods. Inf. Sci. **282**, 111–135 (2014)
7. Cancer Genome Atlas Network: Comprehensive molecular portraits of human breast tumours. Nature **490**(7418), 61 (2012)
8. Chuang, L.Y., Yang, C.H., Wu, K.C., Yang, C.H.: A hybrid feature selection method for DNA microarray data. Comput. Biol. Med. **41**(4), 228–237 (2011)
9. Cortes, C., Vapnik, V.: Support-vector networks. Mach. Learn. **20**(3), 273–297 (1995)
10. Díaz-Uriarte, R., De Andres, S.A.: Gene selection and classification of microarray data using random forest. BMC Bioinf. **7**(1), 3 (2006)
11. Duan, K.B., Rajapakse, J.C., Wang, H., Azuaje, F.: Multiple SVM-RFE for gene selection in cancer classification with expression data. IEEE Trans. Nanobiosci. **4**(3), 228–234 (2005)
12. Golub, T.R., et al.: Molecular classification of cancer: class discovery and class prediction by gene expression monitoring. Science **286**(5439), 531–537 (1999)
13. Gordon, G.J., et al.: Translation of microarray data into clinically relevant cancer diagnostic tests using gene expression ratios in lung cancer and mesothelioma. Cancer Res. **62**(17), 4963–4967 (2002)
14. Guyon, I., Weston, J., Barnhill, S., Vapnik, V.: Gene selection for cancer classification using support vector machines. Mach. Learn. **46**(1–3), 389–422 (2002)
15. Leung, Y., Hung, Y.: A multiple-filter-multiple-wrapper approach to gene selection and microarray data classification. IEEE/ACM Trans. Comput. Biol. Bioinf. **7**(1), 108–117 (2010)
16. Liu, B., Cui, Q., Jiang, T., Ma, S.: A combinational feature selection and ensemble neural network method for classification of gene expression data. BMC Bbioinf. **5**(1), 136 (2004)
17. Mladenović, N., Sifaleras, A., Sörensen, K.: Editorial to the special cluster on variable neighborhood search, variants and recent applications. Int. Trans. Oper. Res. **24**(3), 507–508 (2017)
18. Mundra, P.A., Rajapakse, J.C.: SVM-RFE with MRMR filter for gene selection. IEEE Trans. Nanobiosci. **9**(1), 31–37 (2010)
19. Narendra, P.M., Fukunaga, K.: A branch and bound algorithm for feature subset selection. IEEE Trans. Comput. **9**, 917–922 (1977)

20. Petricoin III, E.F., et al.: Use of proteomic patterns in serum to identify ovarian cancer. Lancet **359**(9306), 572–577 (2002)
21. Reunanen, J.: Search strategies. In: Guyon, I., Nikravesh, M., Gunn, S., Zadeh, L.A. (eds.) Feature Extraction, pp. 119–136. Springer, Heidelberg (2006). https://doi.org/10.1007/978-3-540-35488-8_5
22. Rogati, M., Yang, Y.: High-performing feature selection for text classification. In: Proceedings of the 11th ACM International Conference on Information and Knowledge Management, pp. 659–661 (2002)
23. Sifaleras, A., Salhi, S., Brimberg, J. (eds.): ICVNS 2018. LNCS, vol. 11328. Springer, Cham (2019). https://doi.org/10.1007/978-3-030-15843-9
24. Xie, H., Li, J., Zhang, Q., Wang, Y.: Comparison among dimensionality reduction techniques based on random projection for cancer classification. Comput. Biol. Chem. **65**, 165–172 (2016)

Basic VNS for a Variant of the Online Order Batching Problem

Sergio Gil-Borrás[1]([⊠]) [iD], Eduardo G. Pardo[2]([⊠]) [iD], Antonio Alonso-Ayuso[2] [iD], and Abraham Duarte[2] [iD]

[1] Dept. Sistemas Informáticos, Universidad Politécnica de Madrid, Madrid, Spain
sergio.gil@upm.es
[2] Department of Computer Science, Universidad Rey Juan Carlos, Móstoles, Spain
{eduardo.pardo,antonio.alonso,abraham.duarte}@urjc.es

Abstract. The Online Order Batching Problem is a combinatorial optimization problem related to the process of retrieving items within a warehouse. It appears in the context of warehousing, when the warehouse follows an order-batching picking policy, which means that orders are packed together into batches before been collected. Additionally, since this problem is online, orders are arriving to the warehouse continuously, which is usually due to the fact that orders come from an e-commerce platform. The variant of the problem tacked in this paper also considers an additional characteristic: there are multiple pickers available to collect the batches. In this paper we propose several strategies, based on the Variable Neighborhood Search methodology, to tackle the problem and we compare them with the algorithms in the state of the art, using previously referred data sets. Additionally, we test the influence of different routing strategies not used before in the context of this variant.

Keywords: Online Order Batching Problem · Batching · Variable Neighborhood Search · Multiple pickers

1 Introduction

The e-commerce has suffered an explosion in last few years, thousands of products are sold online everyday and this is just the beginning. The increase in the online sales has made companies to development new processes related to their supply chain management, and also to improve/modify the old ones. However, the evolution in the supply chain models is not new, since it has been happening for many years, as it is possible to trace back in the associated

This research was partially funded by the projects: MTM2015-63710-P, RTI2018-094269-B-I00, TIN2015-65460-C2-2-P and PGC2018-095322-B-C22 from Ministerio de Ciencia, Innovación y Universidades (Spain); by Comunidad de Madrid and European Regional Development Fund, grant ref. P2018/TCS-4566; and by Programa Propio de I+D+i de la Universidad Politécnica de Madrid (Programa 466A).

R. Benmansour et al. (Eds.): ICVNS 2019, LNCS 12010, pp. 17–36, 2020.
https://doi.org/10.1007/978-3-030-44932-2_2

literature from the early eighties up to today. Particularly, in the last ten years, the number of papers related to the supply chain management has suffered a significant increase. In this sense, the online commerce has appeared as one of the next steps. In fact, many classical optimization problems have been reformulated taking into consideration online restrictions.

As part of the supply chain, we focus our attention in the activities that happen within a warehouse. More exactly, in the picking process of items. The global objective of the picking activity is mainly related to satisfy the demand of the customers as soon as possible. However, there are also other important issues in which Warehouse Management Systems (WMS) must pay attention to, such as: balance the workload of the workers in the warehouse, satisfy a predefined due date, save energy, or simply reduce the travel time of the pickers when collecting the items.

In this paper, we tackle the Online Order Batching Problem with Multiple Pickers (OOBPMP). In this optimization problem, orders are arriving to the warehouse 24 h a day/7 days a week, so it means that instances of the problem are changing dynamically. The objective of this optimization problem is either minimizing the time used to collect all the items in the orders received, or minimizing the maximum turnover time of any order (i.e., the time that an order remains in the system). In this problem, the picking strategy is based the concept of batch, which stands for a group of orders that are packed together, before start collecting them. Then, all the items in the same batch are collected in a single route. Notice that the orders can not be split into more than one batch. Also, the batches can not exceed a predefined maximum capacity (weight and/or volume restriction). Every batch can be assigned only to one picker, and every picker can not simultaneously collect items from more than one batch. In this sense, the picking strategy falls into the picker-to-part category. Additionally, the OOBPMP takes into consideration the existence of multiple pickers in the warehouse. In this paper, we propose several strategies to construct the batches, to set the priority in which the batches are assigned to the pickers, and to determine a route to collect the items in the same batch. We do not study here the impact of the storage policy, nor the influence of the different distributions of arrival time moments of the orders.

There are different and well-known routing policies for the picker in the literature related to warehousing, which are suitable for this problem. These strategies range from exact to heuristic methods. The performance of each method partially depends on the shape and structure of the warehouse. The exact methods, further than the longer times needed to calculate a route for the problem, are frequently excluded from real scenarios, because many times they create routes difficult to understand and follow by the operators. This difficulty increases as the complexity of the warehouse grows. On the other hand, simple routing heuristics are usually fast to calculate and they produce reasonable good results with routes easy to understand for the pickers.

The rest of the paper is organized as follows: in Sect. 2, we present the state of the art of the Order Batching family of problems and we focus our attention in the OOBPMP. We also review here the most outstanding heuristic routing procedures in the literature. In Sect. 3, we present a new algorithm for tackling the batching task of the considered problem. Section 4 compiles the computational results obtained with the proposed algorithms over some referenced data sets. Finally, our conclusions and future research lines are exposed in Sect. 5.

2   State of the Art

The Order Batching Problem, further than a single optimization problem, can be considered as a family of optimization problems which groups together those problems related to the retrieval of goods from a warehouse, using a picking policy based on the order batching strategy.

However, within this family of problems, the simplest and most classical version is also referred to in the literature as Order Batching Problem (OBP) [81]. The OBP consists in minimizing the total time needed to collect a group of orders received in a warehouse in a context with a single picker, and having all the orders considered at hand, before starting the batching process. This version of the problem can be considered as static and it has raised a relevant interest in the scientific community. Theoretical studies about the OBP indicated that the problem is $\mathcal{NP}$-hard for general instances [23], but solvable in polynomial time if each batch does not contain more than two orders [23]. However, most of the real instances does not usually fulfill the previous requirement. Due to its hardness, but also to the necessity of finding solutions to the problem in short amounts of time, heuristics and metaheuristics have been applied to tackle the problem. The First-Come First-Served (FCFS) strategy was one of the first heuristic strategies proposed and used in practice to assign orders to batches in a warehouse. This strategy has been widely used due to its simplicity. Other important heuristic methods are the *seed methods* [25,38,62] and the *saving methods* [76]. In [13] it is possible to find a survey of those methods where the authors proposed a classification. The first metaheuristic algorithm applied to the simple OBP was based in a Genetic Algorithm and it was proposed in [42]. Later, a method based on the Variable Neighborhood Search methodology was presented in [1]; an Iterated Local Search in [36] and a Tabu Search in [34]. In [59], the authors proposed an new Iterated Local Search algorithm with a Tabu Thresholding. The current state of the art for the problem, as far as we know, was a multi-start Variable Neighborhood Search method proposed in [53].

Despite of the fact that the simplest OBP has received the largest attention, other static variants have also been studied in the literature: the Order Batching and Sequencing Problem (OBSP) is a variant of the OBP which introduce due dates in the orders [33,52]; and the Min-Max Order Batching Problem (Min-Max OBP) looks for a work balance among several operators in a warehouse [22,55].

As far as the online variants are concerned (i.e., those which receive orders continuously in the system) the first approach found was presented in [81], where the authors proposed a simple FCFS algorithm and considered a variant of the OOBP with multiple pickers. Later, in [86] the OOBP was studied for multiple-block warehouses. In [77] the Online Rescheduling Problem with multiple pickers was tackled by using a Steepest Descent Insertion strategy, and a Multistage Rescheduling strategy. In [27,32,66] a single-block warehouse with a single picker version was tackled in the online context. In the first case the authors proposed an Iterated Local Search and, in the second case, they proposed an Estimation of Distribution Algorithm (EDA). In [94,95] the authors added a new constraint to the problem, related to the scheduling of the delivery. The first work considered only one picker, meanwhile the second one considered multiple pickers. The most recent approach within this context was presented in [10], where the OOBP was studied for multiple blocks and multiple pickers.

We have summarized all the aforementioned methods in Table 1, where the papers are classified depending on the variant of the problem considered. Particularly, we have divided the works into two columns: offline (static) and online (dynamic). For each column we have separated those works which consider only one picker from those which consider multiple pickers. Also we have classified the papers depending on the inclusion or not of due dates in the orders.

Table 1. Publications related with the Order Batching Problem, classified according to the variant of the problem tackled.

		Online	Offline
One picker	With due date	[19]	[4,8,36,43,44,52,82,99]
	Without due date	[11,26,32,45,48,66, 69,72,81,86,87,91, 95,96]	[1,2,5,6,9,13,20,23–25,34,38–40,42,46,49–51,53,54,56,57,59–63, 65,67,68,73,76,78,83–85,88–90,92,93,97,98]
Multiple pickers	With due date	–	[37,41,79,80]
	Without due date	[10,21,77,94]	[3,7,22,35,55]

In this paper we focus our attention in the online version of the OBP which considers multiple pickers and do not include due dates, previously referred to as OOBPMP. Next, in Sect. 2.1 we review the latest batching strategy published for the problem in [94]. Finally, in Sect. 2.2, we review some the most outstanding routing strategies in the context of the OBP.

2.1   Batching State-of-the-Art Algorithm for the OOBPMP

As far as we know, the latest batching algorithm proposed for the OOBPMP was introduced in [94]. The authors used the well-known "seed" strategy [38] for clustering, in order to perform the batching task of the problem. This clustering strategy consists in selecting a "seed" (in this case the seed is represented by an order) as a centroid of a cluster (in this case the cluster is represented by a batch). Then, the seed is assigned to an empty cluster (i.e., the order is assigned to an empty batch) and other available orders might be added to the same batch, depending on the similarity with respect the selected seed order, while the capacity of the batch is not exceeded.

Therefore, for each "seed method" it is necessary to decide how to choose the seed order, and how to determine the similarity of the orders with respect to the seed. In this case, the strategy used to select an order as a "seed" is based on the Smallest Arrival Time rule (i.e., the order which arrived first to the system and has not been assigned yet to any batch is selected as a seed). Once the seed order has been chosen, it is assigned to an empty batch. Then, the strategy used to aggregate other orders to the same batch follows an Aisle-Time-Based strategy. This strategy takes into consideration two dimensions: the percentage of orders that the seed order has in common with the candidate order; and also, since we are in an online context, it includes a measure related to the arrival time of the considered order. This similarity measure is calculated for every order whose device-capacity demand do not exceed the remaining capacity of the picking cart, and then the next order be added to the current batch is selected in a greedy way. Once the batch is full (i.e., no other order among the available ones can be added) the method selects a new seed and so on, until all the orders have been assigned to a batch. We invite the reader to carefully review this method in [94].

2.2   Routing Algorithms

The routing algorithm is responsible for generating a path to collect every item in the orders of the batch, following a single route. As it was aforementioned, the order picking operations are one of the most important and costly processes in a warehouse [12, 16]. In this case, when the picking policy is based on batches, considering the batching and picking tasks together might suppose a reduction up to 35% in the total travel time [14].

The problem of finding a route within the warehouse, where the picker must visit a group of positions, is a simplified version of the Travelling Salesman Problem (TSP) [15] and therefore, there are many different proposals available in the literature to solve it. Particularly, in this case, specific algorithms have been developed considering the rectangular structure of the warehouse used in this paper, which defines a special metric space, whose properties can be used in the design of the route. In fact, there are an exact method [74], based on dynamic programming, which generates the optimal path within this context. However, further than the extra time needed to compute the route, the exact method generates complicated paths for the pickers, which make them to have difficulties to remember and interpret the generated routes [23]. In this sense, other simpler methods, like heuristic ones, are commonly used in practice to solve the problem. The heuristic procedures are usually very fast to compute and they generate simple paths, which are easy to follow for the pickers. Many routing heuristics have been proposed in the literature (see [28,70,71,75] for several proposals and comparisons). In this paper we review the most important and used heuristic procedures in the literature: S-Shape, Largest Gap and Combined.

S-Shape. The S-shape algorithm is one of the most used routing algorithms in the literature mainly due to its simplicity. It is not only easy to implement but also it generates simple routes for the operators. The method constructs a route, starting from the depot, which begins traversing the leftmost aisle which contains at least one item to collect. Then it goes through all the aisles that have items to pick up from any order in the batch. Therefore the picker is performing changes from the front-cross aisle to the back-cross aisle and the other way round. If the number of aisles to be traversed is even, the last parallel aisle will be completely covered (i.e., the picker will finish the route in the front-cross aisle). However, if the number of aisles to traverse is odd, the last corridor is only covered until the last element to be collected and then, the picker performs a U-turn, in order to come back to the front-cross aisle (which contains the depot). An example of a route following this strategy is depicted in Fig. 1. Notice, that in this example there are 5 aisles which contain items to collect.

Fig. 1. Path created with S-Shape method.

Largest Gap. The Largest gap algorithm, along with the S-shape, is one of the routing algorithms widely used in the literature. To understand the largest gap, we first define the concept of gap as the space in an aisle between every two positions which contain an item to collect. Additionally, the space between the front cross-aisle and the first position which contains item to collect and, the space between the last position which contains an item to collect and the back-cross aisle are also considered gaps. For each aisle that has items to pick up, the largest gap in an aisle is the longest distance among all the possible gaps in the aisle. Then, the Largest gap strategy avoids traversing the largest gap of each aisle by performing an U-turn each time a picker arises the position where the largest gap starts/ends. This algorithm also starts exploring the first aisle to the left which contains items to collect. This aisle will be fully traversed, in order to start collecting from the back-cross aisle. Similarly, the last parallel aisle with items to collect will be also fully traversed in order to come back to the front-cross aisle. An example of a route following this strategy is depicted in Fig. 2.

Fig. 2. Path created with Largest Gap method.

Combined. The Combined algorithm was first proposed in [47]. The idea was to combine the two previously introduced methods (S-shape and Largest gap) in order to make a more efficient method. In this case, the algorithm decides, for each parallel aisle, if it is shorter to collect the items in that aisle using an S-shape strategy or a Largest gap one. Then, the algorithm selects the most convenient way. The method has to consider that the number of parallel aisles traversed with S-shape must be even. In some occasions, the circumstances may force the algorithm to change the previously selected strategy for a particular aisle, in order to end the route in the front-cross aisle. An example of a route following this strategy is depicted in Fig. 3.

Fig. 3. Path created with Combined method.

3   Algorithmic Proposal

In this section we present our algorithmic proposal to tackle the OOBPMP. In particular, we propose the use of the Basic Variable Neighborhood Search (BVNS) schema [31,58]. BVNS is a variant of the VNS methodology which was proposed in [58] as a general method to solve hard optimization problems. It is based on the concept of change of the neighborhood structure in order to escape from local minima. There are many different variants of VNS, however the best known ones are: Reduced VNS (RVNS), Basic VNS (BVNS), Variable Neighborhood Descent (VND), General VNS (GVNS), Skewed VNS (SVNS), and Variable Neighborhood Decomposition Search (VNDS). Those variants differs in the use of stochastic/deterministic explorations or a mix of both (as it is the case of BVNS) of the neighborhoods considered. We refer the reader to [29,30,58] for a deep understanding. Some other interesting variants of the VNS methodology have been recently proposed. Among others, we can find: Variable Formulation Search (VFS) [64], Multi-Objective Variable Neighborhood Search [17], and Parallel Variable Neighborhood Search [18,55].

In Algorithm 1 we present a pseudocode of the BVNS method proposed in this paper. The method receives three input parameters: (i) an initial solution S; (ii) a value k_{max}; and (iii) the maximum time (t_{max}). The initial solution will be calculated using an external method that will be described in Sect. 3.1.

On the other hand k_{max} determines the maximum number of neighborhoods that will be explored. Particularly, the method explores the current neighborhood of the solution trying to obtain a better solution. BVNS includes three different stages to explore the current neighborhood and it determines if there has been an improvement. First, the method performs a perturbation to the current solution, in order to escape from the current local optimum. Second, the method runs an improvement procedure based on a local search, which is able to find a local optimum in the current neighborhood. Third, the procedure Neighborhoodchange, determines if there has been any improvement in the solution. If so, the next neighborhood to explore will be the first one. Otherwise, the value of the variable k is increased and, therefore, in the next iteration the number of perturbations performed to the current solution in the Shake procedure is increased. The method stops when the value of k equals k_{max}, and the maximum allowed time is reached.

Algorithm 1. BVNS$(S, k_{\max}, t_{\max})$

1: **repeat**
2:     $k \leftarrow 1$
3:     **while** $k \leq k_{\max}$ **do**
4:         $S' \leftarrow$ Shake(S, k)
5:         $S'' \leftarrow$ LocalSearch(S')
6:         $k \leftarrow$ NeighborhoodChange(S, S'', k)
7:     **end while**
8: **until** $t < t_{max}$
9: **return** S

The description of the Shake and LocalSearch procedures are presented, in Sects. 3.2 and 3.3 respectively. The NeighborhoodChange procedure follows an standard implementation which can be reviewed in [58].

3.1   Constructive Procedure

We have used a random algorithm as a constructive method in order to provide an initial solution to the BVNS algorithm. The constructive algorithm receives a list of orders L_{orders} as an input parameter. In each iteration, an order is randomly selected from the list and it is placed in the next available batch. When the selected order does not fit in the current batch, a new batch is created with this order. Then, the next order will be placed in this new batch and the process is repeated until the order list is fully scanned and all the orders have a batch assigned. Once the process is finished, the procedure returns a list of batches S as a solution. In Algorithm 2 we present a pseudocode of this procedure.

Algorithm 2. Constructive(L_{orders})

1: $S \leftarrow$ NewBatchList()
2: $B \leftarrow$ NewBatch()
3: **repeat**
4:      $o \leftarrow$ ChooseRandomOrder(L_{orders})
5:      $L_{orders} \leftarrow L_{orders} \setminus o$
6:      **if** Fits(B, o) **then**
7:          Add(B, o)
8:      **else**
9:          Add(S, B)
10:         $B \leftarrow$ NewBatch()
11:         Add(B, o)
12:     **end if**
13: **until** $L_{orders} = \emptyset$
14: **return** S

3.2 Shake Procedure

The shake procedure is in charge of performing a perturbation to the current solution. The method starts by selecting two random batches. Then, it selects two random orders (one from each batch) and finally, the method tries to perform an exchange move. The move is not done if the size of any of the selected batches is exceeded. This process is repeated as many times as indicates k.

The Shake procedure receives two input parameters: an initial solution S and the size of the perturbation k. In each iteration, k indicates the number of perturbations to perform. At the end of this procedure, a solution in a different neighborhood is returned. We present a pseudocode of this procedure in Algorithm 3.

Algorithm 3. Shake(S, k)

1: **repeat**
2:      **repeat**
3:          $B_i \leftarrow$ ChooseRandomBatch(S)
4:          $B_j \leftarrow$ ChooseRandomBatch(S)
5:      **until** $B_i \neq B_j$
6:      $o_i \leftarrow$ ChooseRandomOrder(B_i)
7:      $o_j \leftarrow$ ChooseRandomOrder(B_j)
8:      **if** Fits($B_i \setminus o_i, o_j$) **and** Fits($B_j \setminus o_j, o_i$) **then**
9:          $B_i \leftarrow B_i \setminus o_i$
10:         Add(B_i, o_j)
11:         $B_j \leftarrow B_j \setminus o_j$
12:         Add(B_j, o_i)
13:         $k \leftarrow k - 1$
14:     **end if**
15: **until** $k = 0$
16: **return** S

3.3   Local Search Procedure

The BVNS uses a local search in order to deterministically find a local optimum in the current neighborhood. The local search proposed here is based in the one-to-one exchange move. The only input parameter to the local search is a solution S. The method will return another solution which is locally optimum with respect to the initial solution and the neighborhood defined by the exchange move. The local search ends when all candidate interchanges have been explored and no one produces an improve in the current solution. We present the pseudocode of the local search procedure in Algorithm 4.

Algorithm 4. LocalSearch(S)

1: **repeat**
2:      $improved \leftarrow false$
3:      **for** $\forall o_i \in S$ **do**
4:          **for** $\forall o_j \in S$ **do**
5:              $S' \leftarrow$ **Exchange**(S, o_i, o_j)
6:              **if** $f(S') < f(S)$ **then**
7:                  $S \leftarrow S'$
8:                  $improved \leftarrow true$
9:                  **break**
10:              **end if**
11:          **end for**
12:      **end for**
13: **until** $improved = false$
14: **return** S

---

4   Results

In order to test our proposals, we compare our BVNS with the Seed method introduced in [94] and described in Sect. 2.1. The objective function used to compare the algorithms is the total time needed to collect all the orders in a context where the number of pickers is two. The experiments were run an Intel (R) Core (TM) 2 Quad CPU Q6600 2.4 Ghz machine, with 4 GB DDR2 RAM memory. The operating system used was Ubuntu 18.04.1 64 bit LTS, and all the codes were developed in Java 8.

The 64 instances used in our experiments were derived from those reported in [32]. Those instances represent a real warehouse with one block, rectangular shape and 900 storage positions. Particularly, there are 10 aisles with shelves at both sides of the aisle and 45 picking positions in each side. The warehouse layout and the distribution of the items are key elements in the design of the batches. We consider both: random and ABC sorting strategies of the items. The depot (i.e., the place where the items are handed once they have been collected) is placed in the front cross-aisle, either in the left corner or in center of the aisle.

As far as the orders are concerned, the instances contain different number of orders (40, 60, 80, 100). Also there are several picking-cart sizes (30, 45, 60, 75). It is important to notice that, not all the orders represented in each instance are available at the beginning of the process, but they arrive through the working time observed. Particularly, in this paper we consider a time horizon of 4 h. This means that we will observe the behaviour of our algorithms in four hours (remember that in an online problem, the system is actually working 24 h). Then, then number of orders of each instance must arrive to the system in four hours. We use an exponencial distribution for determining the instant in the time when each order arrives to the system, as it is customary in this kind of scenarios.

The BVNS algorithm was parameterized with $k_{max} = 5$ and $t_{max} = 30$ s, and then it has been successfully compared with a variant of the Seed algorithm described in the Sect. 2 using different routing methods. In Table 2 we report the results obtained when using the S-Shape routing method. Similarly, in Table 3 and Table 4 we report the results obtained when using the Largest Gap and Combined routing methods, respectively. For each table, we report the average time used to collect all the orders (Avg. (s)), the deviation with respect to the best value found in the experiment (Dev. (%)) and the number of best solutions found (#Best). In these three experiments we have compared both methods (BVNS and Seed) using the same routing algorithm. Therefore, the results obtained are merit just from the batching strategy. As it is possible to see observing these three tables, BVNS is consistently better than Seed considering both: deviation and number of best solutions found. However, the differences in deviation are very small for the three routing methods. Additionally, when using the Largest Gap routing method, the number of best solutions found by the BVNS and Seed method are almost the same.

However, despite of the fact that we have paired either BVNS and Seed methods with three routing strategies, the original proposal of the Seed method introduced in [94] was based only on the S-Shape routing algorithm. Next, we compare the results obtained by our BVNS paired with Largest Gap and Combined routing methods with respect to the Seed method paired with S-Shape (as it is described by the authors). The results are reported in Tables 5 and 6 respectively. In both cases, the deviation obtained has been considerably improved with respect to the method in the state of the art. Similarly, the number of best-known values has also been increased.

In order to facilitate future comparisons, in the Appendix A we report the best values found per each of the instances considered in this paper.

Table 2. Comparison with the state of the art using the S-Shape routing method.

Batching	BVNS	Seed [94]
Routing	S-Shape	S-Shape
Avg. (s)	32886	33046
Dev. (%)	0,29%	0,91%
#Best	48	28

Table 3. Comparison with the state of the art using the Largest Gap routing method.

Batching	BVNS	Seed [94]
Routing	Largest Gap	Largest Gap
Avg. (s)	32315	32309
Dev. (%)	0,44%	0,51%
#Best	37	36

Table 4. Comparison with the state of the art using the Combined routing method.

Batching	BVNS	Seed [94]
Routing	Combined	Combined
Avg. (s)	31420	31534
Dev. (%)	0,21%	0,66%
#Best	46	24

Table 5. Comparison between the BVNS paired with Largest Gap with respect to the Seed method paired with S-Shape.

Batching	BVNS	Seed [94]
Routing	Largest Gap	S-Shape
Avg. (s)	32315	33046
Dev. (%)	0,91%	3,18%
#Best	42	22

Table 6. Comparison between the BVNS paired with Combined with respect to the Seed method paired with S-Shape.

Batching	BVNS	Seed [94]
Routing	Combined	S-Shape
Avg. (s)	31420	33046
Dev. (%)	0,04%	5,01%
#Best	60	4

5   Conclusions

In this paper we have dealt with a variant of the Online Order Batching Problem. Particularly, the variant which considers multiple pickers to collect the items in the batches. We have reviewed the state of the art of the problem and highlighted the latest approach to tackle it. We have also designed an algorithm, based on the Basic Variable Neighborhood Search methodology, in order to provide good quality solutions for the OOBPMP. The obtained results have been compared with the state of the art using different routing algorithms. In all the considered cases, the BVNS proposed improved the previous method in the state of the art. We also noticed that the use of the Combined routing method was the most effective among the considered ones for this problem.

A   Best-Known Values per Instance

See Table 7.

Table 7. Best-known values of the objective function Time (s) per instance.

Instance	Time (s)	Instance	Time (s)	Instance	Time (s)
abc1_40_29	23992	abc1_40_30	22398	abc1_40_31	22526
abc1_40_32	22528	abc1_60_37	31029	abc1_60_38	28469
abc1_60_39	25501	abc1_60_40	24796	abc1_80_61	41391
abc1_80_62	33856	abc1_80_63	29225	abc1_80_64	29963
abc1_100_69	44026	abc1_100_70	35298	abc1_100_71	32068
abc1_100_72	31089	abc2_40_9	23942	abc2_40_10	21370
abc2_40_11	21098	abc2_40_12	20909	abc2_60_17	29979
abc2_60_18	29193	abc2_60_19	27954	abc2_60_20	22775
abc2_80_45	38246	abc2_80_46	31943	abc2_80_47	30175
abc2_80_48	28280	abc2_100_53	49381	abc2_100_54	35574
abc2_100_55	36281	abc2_100_56	30630	ran1_40_29	26011
ran1_40_30	23601	ran1_40_31	24904	ran1_40_32	21948
ran1_60_37	35830	ran1_60_38	30448	ran1_60_39	26893
ran1_60_40	27907	ran1_80_61	47704	ran1_80_62	39016
ran1_80_63	31755	ran1_80_64	30074	ran1_100_69	52084
ran1_100_70	39676	ran1_100_71	35887	ran1_100_72	33779
ran2_40_9	25476	ran2_40_10	25794	ran2_40_11	22076
ran2_40_12	21629	ran2_60_17	33661	ran2_60_18	30619
ran2_60_19	26974	ran2_60_20	26496	ran2_80_45	44121
ran2_80_46	36148	ran2_80_47	32117	ran2_80_48	30286
ran2_100_53	56735	ran2_100_54	40345	ran2_100_55	40688
ran2_100_56	33380				

References

1. Albareda-Sambola, M., Alonso-Ayuso, A., Molina, E., De Blas, C.S.: Variable neighborhood search for order batching in a warehouse. Asia Pac. J. Oper. Res. **26**(5), 655–683 (2009)

2. Ardjmand, E., Bajgiran, O.S., Youssef, E.: Using list-based simulated annealing and genetic algorithm for order batching and picker routing in put wall based picking systems. Appl. Soft Comput. **75**, 106–119 (2019)

3. Ardjmand, E., Shakeri, H., Singh, M., Bajgiran, O.S.: Minimizing order picking makespan with multiple pickers in a wave picking warehouse. Int. J. Prod. Econ. **206**, 169–183 (2018)

4. Azadnia, A.H., Taheri, S., Ghadimi, P., Mat Saman, M.Z., Wong, K.Y.: Order batching in warehouses by minimizing total tardiness: a hybrid approach of weighted association rule mining and genetic algorithms. Sci. World J. **2013** (2013)

5. Bozer, Y.A., Kile, J.W.: Order batching in walk-and-pick order picking systems. Int. J. Prod. Res. **46**(7), 1887–1909 (2008)

6. Briant, O., Cambazard, H., Cattaruzza, D., Catusse, N., Ladier, A.L., Ogier, M.: A column generation based approach for the joint order batching and picker routing problem. In: ROADEF 2018 (2018)

7. Bué, M., Cattaruzza, D., Ogier, M., Semet, F.: An integrated order batching and picker routing problem. HAL (hal-01849980) (2018)

8. Bustillo, M., Menéndez, B., Pardo, E.G., Duarte, A.: An algorithm for batching, sequencing and picking operations in a warehouse. In: 2015 International Conference on Industrial Engineering and Systems Management (IESM), pp. 842–849, October 2015

9. Cano, J.A., Correa-Espinal, A.A., Gómez-Montoya, R.A.: Solución del problema de conformación de lotes en almacenes utilizando algoritmos genéticos. Información tecnológica **29**(6), 235–244 (2018)

10. Chen, F., Wei, Y., Wang, H.: A heuristic based batching and assigning method for online customer orders. Flex. Serv. Manuf. J. **30**(4), 640–685 (2018). https://doi. org/10.1007/s10696-017-9277-7

11. Chew, E.P., Tang, L.C.: Travel time analysis for general item location assignment in a rectangular warehouse. Eur. J. Oper. Res. **112**(3), 582–597 (1999)

12. Coyle, J.J., Bardi, E.J., Langley, C.J., et al.: The Management of Business Logistics, vol. 6. West Publishing Company Minneapolis, St Paul (1996)

13. De Koster, M.B.M., Van der Poort, E.S., Wolters, M.: Efficient order batching methods in warehouses. Int. J. Prod. Res. **37**(7), 1479–1504 (1999)

14. de Koster, R., Roodbergen, K.J., van Voorden, R.: Reduction of walking time in the distribution center of De Bijenkorf. In: Speranza, M.G., Stähly, P. (eds.) New Trends in Distribution Logistics. Lecture Notes in Economics and Mathematical Systems, vol. 480, pp. 215–234. Springer, Heidelberg (1999). https://doi.org/10. 1007/978-3-642-58568-5_11

15. De Koster, R., Le-Duc, T., Roodbergen, K.J.: Design and control of warehouse order picking: a literature review. Eur. J. Oper. Res. **182**(2), 481–501 (2007)

16. Drury, J.: Towards more efficient order picking. IMM Monograph, No. 1 (1988)

17. Duarte, A., Pantrigo, J.J., Pardo, E.G., Mladenovic, N.: Multi-objective variable neighborhood search: an application to combinatorial optimization problems. J. Glob. Optim. **63**(3), 515–536 (2015). https://doi.org/10.1007/s10898-014-0213-z

18. Duarte, A., Pantrigo, J.J., Pardo, E.G., Sánchez-Oro, J.: Parallel variable neighbourhood search strategies for the cutwidth minimization problem. IMA J. Manag. Math. **27**(1), 55–73 (2016)

19. Elsayed, E., Lee, M.K.: Order processing in automated storage/retrieval systems with due dates. IIE Trans. **28**(7), 567–577 (1996)

20. Elsayed, E.A.: Algorithms for optimal material handling in automatic warehousing systems. Int. J. Prod. Res. **19**(5), 525–535 (1981)

21. van der Gaast, J.P., Jargalsaikhan, B., Roodbergen, K.J.: Dynamic batching for order picking in warehouses. In: 15th IMHRC Proceedings, Savannah, Georgia, USA (2018)
22. Gademann, A.J.R.M., Van Den Berg, J.P., Van Der Hoff, H.H.: An order batching algorithm for wave picking in a parallel-aisle warehouse. IIE Trans. **33**(5), 385–398 (2001)
23. Gademann, N., Velde, S.: Order batching to minimize total travel time in a parallel-aisle warehouse. IIE Trans. **37**(1), 63–75 (2005)
24. Galka, S., Ulbrich, A., Günthner, W.: Performance calculation for order picking systems by analytical methods and simulation. Technical report, Technische Universität München, München, Germany (2008)
25. Gibson, D.R., Sharp, G.P.: Order batching procedures. Eur. J. Oper. Res. **58**(1), 57–67 (1992)
26. Gil-Borrás, S., Duarte, A., Alonso-Ayuso, A., Pardo, E.G.: Búsqueda de vecindad variable para el problema de la agrupación y recogida de pedidos online en almacenes logísticos. In: XVIII Conferencia de la Asociación Española para la Inteligencia Artificial, Granada, España, pp. 551–556, October 2018
27. Gil-Borrás, S., Pardo, E.G., Alonso-Ayuso, A., Duarte, A.: New VNS variants for the online order batching problem. In: Sifaleras, A., Salhi, S., Brimberg, J. (eds.) ICVNS 2018. LNCS, vol. 11328, pp. 89–100. Springer, Cham (2019). https://doi.org/10.1007/978-3-030-15843-9_8
28. Hall, R.W.: Distance approximations for routing manual pickers in a warehouse. IIE Trans. **25**(4), 76–87 (1993)
29. Hansen, P., Mladenović, N.: Variable neighborhood search: principles and applications. Eur. J. Oper. Res. **130**(3), 449–467 (2001)
30. Hansen, P., Mladenović, N., Moreno-Pérez, J.A.: Variable neighbourhood search: methods and applications. Ann. Oper. Res. **175**(1), 367–407 (2010). https://doi.org/10.1007/s10479-009-0657-6
31. Hansen, P., Mladenović, N., Todosijević, R., Hanafi, S.: Variable neighborhood search: basics and variants. EURO J. Comput. Optim. **5**(3), 423–454 (2017). https://doi.org/10.1007/s13675-016-0075-x
32. Henn, S.: Algorithms for on-line order batching in an order picking warehouse. Comput. Oper. Res. **39**(11), 2549–2563 (2012)
33. Henn, S., Schmid, V.: Metaheuristics for order batching and sequencing in manual order picking systems. Comput. Ind. Eng. **66**(2), 338–351 (2013)
34. Henn, S., Wäscher, G.: Tabu search heuristics for the order batching problem in manual order picking systems. Eur. J. Oper. Res. **222**(3), 484–494 (2012)
35. Henn, S.: Order batching and sequencing for the minimization of the total tardiness in picker-to-part warehouses. Flex. Serv. Manuf. J. **27**(1), 86–114 (2015). https://doi.org/10.1007/s10696-012-9164-1
36. Henn, S., Koch, S., Doerner, K.F., Strauss, C., Wäscher, G.: Metaheuristics for the order batching problem in manual order picking systems. Bus. Res. **3**(1), 82–105 (2010)
37. Henn, S., et al.: Variable neighborhood search for the order batching and sequencing problem with multiple pickers. Technical report, Otto-von-Guericke University Magdeburg, Faculty of Economics and Management (2012)
38. Ho, Y.C., Tseng, Y.Y.: A study on order-batching methods of order-picking in a distribution centre with two cross-aisles. Int. J. Prod. Res. **44**(17), 3391–3417 (2006)
39. Ho, Y.C., Su, T.S., Shi, Z.B.: Order-batching methods for an order-picking warehouse with two cross aisles. Comput. Ind. Eng. **55**(2), 321–347 (2008)

40. Hong, S., Johnson, A.L., Peters, B.A.: Analysis of picker blocking in narrow-aisle batch picking. Technical report, Texas A&M University (2010)
41. Hong, S., Johnson, A.L., Peters, B.A.: Batch picking in narrow-aisle order picking systems with consideration for picker blocking. Eur. J. Oper. Res. **221**(3), 557–570 (2012)
42. Hsu, C.M., Chen, K.Y., Chen, M.C.: Batching orders in warehouses by minimizing travel distance with genetic algorithms. Comput. Ind. **56**(2), 169–178 (2005)
43. Huang, M., Guo, Q., Liu, J., Huang, X.: Mixed model assembly line scheduling approach to order picking problem in online supermarkets. Sustainability **10**(11), 3931 (2018)
44. Jiang, X., Zhou, Y., Zhang, Y., Sun, L., Hu, X.: Order batching and sequencing problem under the pick-and-sort strategy in online supermarkets. Procedia Comput. Sci. **126**, 1985–1993 (2018)
45. Kamin, N.: On-line optimization of order picking in an automated warehouse. Ph.D. thesis, Technische Universität Belin, Belin, Germany (1998)
46. Koch, S., Wäscher, G.: A grouping genetic algorithm for the Order Batching Problem in distribution warehouses. J. Bus. Econ. **86**(1–2), 131–153 (2016). https://doi.org/10.1007/s11573-015-0789-x
47. Koster, R.D., Poort, E.V.D.: Routing orderpickers in a warehouse: a comparison between optimal and heuristic solutions. IIE Trans. **30**(5), 469–480 (1998)
48. Le-Duc, T.: Design and control of efficient order picking processes. Ph.D. thesis, Erasmus University Rotterdam. Erasmus Research Institute of Management, Rotterdam, Holland, September 2005
49. Lenoble, N., Frein, Y., Hammami, R.: Optimization of order batching in a picking system with a vertical lift module. In: Temponi, C., Vandaele, N. (eds.) ILS 2016. LNBIP, vol. 262, pp. 153–167. Springer, Cham (2018). https://doi.org/10.1007/978-3-319-73758-4_11
50. Lenoble, N., Frein, Y., Hammami, R.: Optimization of order batching in a picking system with carousels. In: 20th World Congress of the International Federation of Automatic Control, IFAC 2017 (2017)
51. Lin, C.C., Kang, J.R., Hou, C.C., Cheng, C.Y.: Joint order batching and picker manhattan routing problem. Comput. Ind. Eng. **95**, 164–174 (2016)
52. Menéndez, B., Bustillo, M., Pardo, E.G., Duarte, A.: General variable neighborhood search for the order batching and sequencing problem. Eur. J. Oper. Res. **263**(1), 82–93 (2017)
53. Menéndez, B., Pardo, E.G., Alonso-Ayuso, A., Molina, E., Duarte, A.: Variable neighborhood search strategies for the order batching problem. Comput. Oper. Res. **78**, 500–512 (2017)
54. Menéndez, B., Pardo, E.G., Duarte, A., Alonso-Ayuso, A., Molina, E.: General variable neighborhood search applied to the picking process in a warehouse. Electron. Notes Discrete Math. **47**, 77–84 (2015)
55. Menéndez, B., Pardo, E.G., Sánchez-Oro, J., Duarte, A.: Parallel variable neighborhood search for the min-max order batching problem. Int. Trans. Oper. Res. **24**(3), 635–662 (2017)
56. Menéndez, B., Pardo, E.G., Duarte, A.: Búsqueda de vecindad variable general aplicada al proceso de recogida de productos en almacenes. In: XVI Conferencia de la Asociación Española para la Inteligencia Artificial, Albacete, España, Noviembre 2015
57. Miguel, F., Frutos, M., Tohmé, F., Rossit, D.: A memetic algorithm for the integral obp/opp problem in a logistics distribution center. Uncertain Supply Chain. Manag. **7**(2), 203–214 (2019)

58. Mladenović, N., Hansen, P.: Variable neighborhood search. Comput. Oper. Res. **24**(11), 1097–1100 (1997)
59. Öncan, T.: MILP formulations and an iterated local search algorithm with Tabu thresholding for the order batching problem. Eur. J. Oper. Res. **243**(1), 142–155 (2015)
60. Öncan, T., Cağırıcı, M.: MILP formulations for the order batching problem in low-level picker-to-part warehouse systems. IFAC Proc. Vol. **46**(9), 471–476 (2013)
61. Oncan, T.: A genetic algorithm for the order batching problem in low-level picker-to-part warehouse systems. In: Proceedings of the International MultiConference of Engineers and Computer Scientists, vol. 1 (2013)
62. Pan, C.H., Liu, S.Y.: A comparative study of order batching algorithms. Omega **23**(6), 691–700 (1995)
63. Pan, J.C.H., Shih, P.H., Wu, M.H.: Order batching in a pick-and-pass warehousing system with group genetic algorithm. Omega **57**, 238–248 (2015)
64. Pardo, E.G., Mladenović, N., Pantrigo, J.J., Duarte, A.: Variable formulation search for the cutwidth minimization problem. Appl. Soft Comput. **13**(5), 2242–2252 (2013)
65. Parikh, P.J.: Designing order picking systems for distribution centers. Ph.D. thesis, Virginia Tech. Faculty of the Virginia Polytechnic Institute and State University, Blacksburg, Virginia, USA (2006)
66. Pérez-Rodríguez, R., Hernández-Aguirre, A., Jöns, S.: A continuous estimation of distribution algorithm for the online order-batching problem. Int. J. Adv. Manuf. Technol. **79**(1), 569–588 (2015). https://doi.org/10.1007/s00170-015-6835-6
67. Pérez-Rodríguez, R., Hernández-Aguirre, A.: An estimation of distribution algorithm-based approach for the order batching problem. Res. Comput. Sci. **93**, 141–150 (2015)
68. Pérez-Rodríguez, R., Hernández-Aguirre, A.: An estimation of distribution algorithm-based approach for the order batching problem: an experimental study. In: Handbook of Research on Military, Aeronautical, and Maritime Logistics and Operations, pp. 509–518. IGI Global (2016)
69. Pérez-Rodríguez, R., Hernández-Aguirre, A.: Finding interactions or relationships between customer orders for building better batches by means of an estimation of distribution algorithm-based approach for the online order batching problem. In: Proceedings of the Genetic and Evolutionary Computation Conference 2016, pp. 989–996. ACM (2016)
70. Petersen, C.G.: An evaluation of order picking routeing policies. Int. J. Oper. Prod. Manag. **17**(11), 1098–1111 (1997)
71. Petersen, C.: Routeing and storage policy interaction in order picking operations. Decis. Sci. Inst. Proc. **31**(3), 1614–1616 (1995)
72. Postema, J.T.: Metaheuristics for order batching in ecommerce warehouses. In: 27th Twente Student Conference on IT. University of Twente, Faculty of Electrical Engineering, Mathematics and Computer Science, Enschede, The Netherlands, July 2017
73. Raj, L.S., Girubha, R.J.: Aggregation of order picking system using order batching. IOSR J. Mech. Civ. Eng. **11**(2), 01–04 (2014)
74. Ratliff, H.D., Rosenthal, A.S.: Order-picking in a rectangular warehouse: a solvable case of the traveling salesman problem. Oper. Res. **31**(3), 507–521 (1983)
75. Roodbergen, K.J., Petersen, C.G.: How to improve order picking efficiency with routing and storage policies. In: Progress in Material Handling Practice, pp. 107–124 (1999)

76. Rosenwein, M.B.: A comparison of heuristics for the problem of batching orders for warehouse selection. Int. J. Prod. Res. **34**(3), 657–664 (1996)
77. Rubrico, J., Higashi, T., Tamura, H., Ota, J.: Online rescheduling of multiple picking agents for warehouse management. Robot. Comput. Integr. Manuf. **27**(1), 62–71 (2011)
78. Scholz, A., Wäscher, G.: Order Batching and Picker Routing in manual order picking systems: the benefits of integrated routing. Cent. Eur. J. Oper. Res. **25**(2), 491–520 (2017). https://doi.org/10.1007/s10100-017-0467-x
79. Scholz, A., Schubert, D., Wäscher, G.: Order picking with multiple pickers and due dates-simultaneous solution of order batching, batch assignment and sequencing, and picker routing problems. Eur. J. Oper. Res. **263**(2), 461–478 (2017)
80. Schubert, D., Scholz, A., Wäscher, G.: Integrated order picking and vehicle routing with due dates. OR Spectr. **40**(4), 1109–1139 (2018). https://doi.org/10.1007/s00291-018-0517-3
81. Tang, L.C., Chew, E.P.: Order picking systems: batching and storage assignment strategies. Comput. Ind. Eng. **33**(3), 817–820 (1997). Selected Papers from the Proceedings of 1996 ICC&IC
82. Tsai, C.Y., Liou, J.J., Huang, T.M.: Using a multiple-ga method to solve the batch picking problem: considering travel distance and order due time. Int. J. Prod. Res. **46**(22), 6533–6555 (2008)
83. Valle, C.A., Beasley, J.E., da Cunha, A.S.: Optimally solving the joint order batching and picker routing problem. Eur. J. Oper. Res. **262**(3), 817–834 (2017)
84. Valle, C.A., Beasley, J.E.: Order batching for picker routing using a distance approximation. arXiv preprint arXiv:1808.00499 (2018)
85. Van Gils, T., Braekers, K., Ramaekers, K., Depaire, B., Caris, A.: Improving order picking efficiency by analyzing the combination of storage, batching, zoning and routing policies in a 2-block warehouse. Technical report, Hasselt University, Martelarenlaan 42, 3500 Hasselt, Belgium (2016)
86. Van Nieuwenhuyse, I., de Koster, R., Colpaert, J.: Order batching in multi-server pick-and-sort warehouses. Katholieke Universiteit Leuven, Department of Decision Sciences and Information Management, vol. 180, no. 140, pp. 367–8869 (2007)
87. Van Nieuwenhuyse, I., de Koster, R.B.: Evaluating order throughput time in 2-block warehouses with time window batching. Int. J. Prod. Econ. **121**(2), 654–664 (2009)
88. Verschure, A.: Improving picking efficiency in a warehouse with multiple floors at Docdata NV. Ph.D. thesis, Technische Universities Eindhoven, Eindhoven, Netherlands (2014)
89. Wäscher, G., Scholz, A., et al.: A solution approach for the joint order batching and picker routing problem in a two-block layout. Technical report, Otto-von-Guericke University Magdeburg, Faculty of Economics and Management (2015)
90. Wasusri, T., Theerawongsathon, P.: An application of discrete event simulation on order picking strategies: A case study of footwear warehouses. In: Claus, T., Frank Herrmann, M.M.O.R. (eds.) Proceedings 30th European Conference on Modelling and Simulation - ECMS, pp. 121–127 (2016)
91. Won, J., Olafsson, S.: Joint order batching and order picking in warehouse operations. Int. J. Prod. Res. **43**(7), 1427–1442 (2005)
92. Won, J.: Order batching and picking optimization in terms of supply chain management. Ph.D. thesis, Iowa State University, Iowa, USA (2004)
93. Yu, M.M.: Enhancing warehouse performance by efficient order picking. Ph.D. thesis, Erasmus University Rotterdam. Erasmus Research Institute of Management, Rotterdam, Holland (2008)

94. Zhang, J., Wang, X., Chan, F.T.S., Ruan, J.: On-line order batching and sequencing problem with multiple pickers: a hybrid rule-based algorithm. Appl. Math. Model. **45**, 271–284 (2017)
95. Zhang, J., Wang, X., Huang, K.: Integrated on-line scheduling of order batching and delivery under B2C e-commerce. Comput. Ind. Eng. **94**, 280–289 (2016)
96. Zhang, J., Wang, X., Huang, K.: On-line scheduling of order picking and delivery with multiple zones and limited vehicle capacity. Omega **79**, 104–115 (2018)
97. Zhu, J., Zhang, H., Zhou, L., Guo, J.: Order batching optimization in dual zone type warehouse based on genetic algorithms. Sci. J. Bus. Manag. **3**(3), 77–81 (2015)
98. Žulj, I., Kramer, S., Schneider, M.: A hybrid of adaptive large neighborhood search and tabu search for the order-batching problem. Eur. J. Oper. Res. **264**(2), 653–664 (2018)
99. Zuniga, C., Olivares-Benitez, E., Tenahua, A., Mujica, M.: A methodology to solve the order batching problem. IFAC-PapersOnLine **48**(3), 1380–1386 (2015)

A VNS-Based Algorithm for the Mammography Unit Location Problem

Marcone Jamilson Freitas Souza$^{(\boxtimes)}$ ⓘ, Puca Huachi Vaz Penna ⓘ,
Manoel Victor Stilpen Moreira de Sá ⓘ, and Patrick Moreira Rosa ⓘ

Departamento de Computação, Universidade Federal de Ouro Preto (UFOP),
Campus Universitário, Morro do Cruzeiro, Ouro Preto, MG 35.400-000, Brazil
{marcone,puca}@ufop.edu.br,
{manoel.stilpen,patrick.moreira}@aluno.ufop.edu.br

Abstract. This work deals with the mammography unit location problem in Brazil. In this problem, there is a set of mammography units to be installed in cities with hospital infrastructure and a set of cities, each with a demand for mammography screenings to be performed in women aged 40 to 69 years old. The goal is to decide where to install mammography units to maximize the total demand, satisfying the constraints that a woman can not travel more than 60 km to be attended and that not all cities are candidates to host a mammography unit. One mathematical programming formulation and a VNS-based algorithm are introduced. The methods were tested using data from Minas Gerais State, Brazil. We analyze the performance of the VNS algorithm considering several scenarios created from the base instance. The results show that the proposed algorithm is able to provide good quality solutions quickly. In addition, it has been shown that with the proposed allocation it is possible to increase the coverage of mammography screenings in the real instance.

Keywords: Mammography unit location · Maximal Covering Location Problem · Variable Neighborhood Search · Mathematical programming

1 Introduction

Among female population, cancer is the second leading cause of death worldwide, accounting for 14% of all deaths. Breast cancer is the most commonly diagnosed cancer among women in most countries of the world [17]. This situation is not different in Brazil [13].

According to [19], the reduction in the number of breast cancer-related deaths in the female population is directly related to the early diagnosis of this disease. On the other hand, the screening by mammography unit is the primary means of early detection of breast cancer.

The current recommendation of the Health Ministry of Brazil is that mammography screenings should be offered to women aged from 50 to 69 years old

R. Benmansour et al. (Eds.): ICVNS 2019, LNCS 12010, pp. 37–52, 2020.
https://doi.org/10.1007/978-3-030-44932-2_3

biennially, as this age group benefits more from the examination in terms of traceability [11,12]. Moreover, studies show an additional 8.9% screenings annually for a diagnostic indication to women in this age group. Thus, for women aged 50–69, the estimated demand is 58.9% of the female population per year [11]. Besides, according to these studies, annual screenings are required in 20% of women between the ages of 40 and 49, of which 10% are for diagnostic purposes and 10% for other indications.

Mammography screening is one of the diagnostic services offered by the Brazilian government health care, named Unified Health System (SUS, in Portuguese), through which a large part of the Brazilian population has its health care needs satisfied. According to the National Cancer Institute [14], each equipment is capable of performing 5,069 mammography screenings annually. Federal Government researches show that 70% of the population have SUS as a reference in health care and use the Health Care Network (RAS) to perform health services under the SUS, among them the diagnostic support services.

Regarding access to health services in RAS, Andrade et al. [4] emphasize that studies are essential to optimize mammography unit allocation. The woman travel distance to the place where the equipment is installed is one of the factors that most contribute to women failing the screening. In other words, many women do not perform the mammography screening simply because the equipment is installed far from their residences.

The SUS inefficiency in offering mammography screenings to the Brazilian female population is verified in several works, as in [3], [4] and [18]. These authors verified that considering only the demand for mammography screenings to be performed annually and the existing number of mammography units, the current number of equipment is sufficient. However, the distribution of this equipment is inadequate, since some regions are well covered and others are not. Besides, in many locations, there is a skilled labor shortage to operate the equipment.

This work deals with the Mammography Unit Location Problem (MULP) and contributes to the development of optimization models for a better distribution of mammography units, while at the same time doing a preliminary case study of Minas Gerais State, Brazil. A heuristic algorithm based on the Variable Neighborhood Search (VNS) method is presented to obtain approximate solutions to the Maximal Covering Location Problem (MCLP) [5]. This heuristic algorithm is proposed since the MULP is NP-hard [7].

The rest of this paper is organized as follows. In Sect. 2, a literature review is made, while in Sect. 3 the problem under study is described. Section 4 introduces a mathematical programming formulation for solving the MULP and Sect. 5 presents a VNS-based algorithm to obtain high quality solutions for it. In Sect. 6, the computational results are reported. Section 7 concludes the work and presents perspectives for future work.

2   Literature Review

In [4], the authors analyzed the number of existing mammography units and the female population that requires mammography screenings in the State of Minas

Gerais in 2012. The authors concluded that distance and women displacement time to the equipment are important limiting factors for mammography screening. According to them, if the equipment is far from the women's residence, it is very likely that they will not travel to realize the mammography screenings. The authors emphasize the importance of studies to optimize these mammography unit allocation.

According to [3], several factors can create barriers to health services accessibility, such as educational level, socioeconomic status, transportation cost, health center location. The concept of accessibility is not only related to the availability of resources in a given period ([1] *apud* [3]); in fact, it is also related to the ability of individuals to appropriate the services offered. According to [8], it is not enough to offer the health service, it is also necessary that the patient can reach the center where it is offered at reasonable times and costs. In [3], the authors conclude that the availability of mammography units in Brazil is sufficient to cover the full demand of women for mammography screenings. However, when the maximum distance restriction is added in the context, the equipment distribution is inadequate since many of them do not cover all regions.

The demands of the public health medical specialties in the Minas Gerais State were studied in [16]. The object of research was the location of 51 Medical Specialty Centers (CEMs) in 853 cities of the State, and in five specialties: cardiology, pediatrics, mastology, gynecology, and endocrinology. They were chosen by the criterion of higher demand for medical attention in the State and medical care hours. The authors proposed a mixed integer programming model, based on the Maximal Covering Location Problem (MCLP), and considered three scenarios to define a set of candidate cities to receive a CEM. The first scenario considered 853 cities as candidates, the second 372, and the third 98. The maximum distance parameter varied in the values 400, 300, 200, and 100 kilometers (km), in order to identify the configuration that provides the highest coverage and the shortest average distance of displacement. The cities distance matrix was obtained by calculating the distance between two points according to the spherical law of the cosines, updated by a correction factor. The authors verified that the selected variations showed a better geographic distribution for the 51 CEMs with smaller distances of maximum coverage in all scenarios. Moreover, given the economic crisis in the Minas Gerais State, they suggested adopting the third scenario, considering the possibility of cost reduction as well as the number of CEMs to be installed, without coverage demand loss.

In [6], the authors analyzed the mammography units location in a set of 12 health regions of Minas Gerais State, involving 142 cities. The authors developed four mathematical programming formulations, all of them based on the p-median problem. In the first one, the goal is to minimize the total distance traveled by women when going to the mammography center. In the second formulation, the maximum displacement distance constraint is relaxed, and the distance exceeding the maximum distance is penalized in the objective function. The last two formulations differ from the previous two because they consider as objective function the distance and the women demand to be attended. More precisely, the objective function is given by the product between the traveled distance to the mammography unit and the number of women who travel. The objective of these last two formulations is to encourage the installation of equipment in cities with the highest mammography demand. As observed in [3], the authors concluded that there are more mammography units in the analyzed region than necessary and the current location is inadequate because it does not comply with the recommended Health Ministry rules.

A comprehensive review of models and solution methods for the healthcare facility location problem can be found in [2].

3   Problem Statement

The Mammography Unit Location Problem (MULP) addressed here has the following characteristics:

(a) There is a set S of n candidate cities to host p mammography units, with $p < n$;
(b) Each mammography unit has an annual capacity of realizing *cap* mammography screenings;
(c) Each city has an annual demand of mammography screenings for women in the age range indicated to do the screening, that is, 58.9% of women aged 50–69 and 20% of women between the ages of 40 and 49, according to the current recommendation of the Health Ministry of Brazil;
(d) A woman cannot travel more than R km to a city that hosts a mammography unit;
(e) Only cities with hospital infrastructure are candidates to host mammography units. In this paper we consider that a city is candidate to host mammography equipment if it has at least *demMin* women in the age range indicated for realizing the screening;
(f) Each city must be either fully covered by a mammography equipment or not covered. That is, we consider in this paper that a city cannot be partially covered. This restriction is imposed for administrative reasons, since this would require managing which women in a city should do the mammography screenings.

The objective is to decide where to install the mammography units in order to maximize the total demand for mammography screenings.

4   Mathematical Formulation

For applying the proposed formulation, we assume that the demand for mammography screenings of each city is smaller than the capacity for screenings of a mammography unit. When this does not happen, we allocate as many mammography units as necessary until the demand is less than the equipment capacity. In this way, the demand covered with this preprocessing is maximum.

In order to introduce the model, the input parameters and the decision variables are defined according to Table 1.

Table 1. Parameters and decision variables

Parameters	
N	Set of cities
d_{ij}	Distance from city i to city j
dem_j	Demand for mammography screenings in city j
cap	Annual screening capacity of an equipment
p	Amount of mammography units to be located
R	Maximum travel distance to be served
$demMin$	Minimum annual screening demand that a city must have to host an equipment
S_i	Set of cities whose distance from city i is less or equal to R km, that is, $S_i = \{j \in N \mid d_{ij} \leq R \text{ and } d_{ji} \leq R\}$
Decision variables	
x_{ij}	Binary variable that assumes value 1 if the women from city j are served by an equipment installed at city i and value 0, otherwise
y_i	Integer variable that represents the number of equipment installed at city i
z_i	Binary variable that assumes value 1 if the city i hosts some equipment and value 0, otherwise

Equations (1) to (11) represent the MULP:

$$\max \quad \sum_{i \in N} \sum_{j \in S_i} dem_j \cdot x_{ij} \tag{1}$$

$$\text{s. t.} \quad \sum_{i \in S_j} x_{ij} \quad \leq \quad 1 \quad \forall j \in N \tag{2}$$

$$\sum_{i \in N} y_i \quad = \quad p \tag{3}$$

$$\sum_{j \in S_i} dem_j \cdot x_{ij} \quad \leq \quad cap \cdot y_i \; \forall i \in N \tag{4}$$

$$z_i \quad \geq \quad y_i/p \quad \forall i \in N \tag{5}$$

$$z_i \quad \geq \quad x_{ij} \quad \forall i, j \in N \tag{6}$$

$$x_{ii} \quad = \quad z_i \quad \forall i \in N \tag{7}$$

$$y_i \quad = \quad 0 \quad \forall i \in N \mid dem_i < demMin \tag{8}$$

$$x_{ij} \quad \in \quad \{0, 1\} \quad \forall i, j \in N \tag{9}$$

$$y_i \quad \in \quad \mathbb{Z}^+ \quad \forall i \in N \tag{10}$$

$$z_i \quad \in \quad \{0, 1\} \quad \forall i \in N \tag{11}$$

The objective function (1) aims to maximize the total demand for mammography screenings. Constraints (2) indicate that each city j must be served by a single mammography unit installed in city i, or not served. Constraint (3) determines that all available p equipment must be allocated, and a city may receive more than one equipment. Constraints (4) guarantee that the equipment' capacity must be respected. Constraints (5) ensure that if at least one equipment is installed in the city i then the variable z_i assumes the value 1. Constraints (6) ensure that a city j can only be served by a city i if an equipment is installed in this city. Constraints (7) ensure that the demand of city i has to be covered by the equipment installed in the city itself. Constraints (8) indicate that an equipment can only be installed in a city that has the demand for mammography screenings greater than or equal to $demMin$, to economically justify its installation. Finally, Constraints (9), (10) and (11) impose the domain of the decision variables.

5   The Proposed VNS Algorithm

As stated in Sect. 4, we assume that there is an initial preprocessing stage to allocate mammography units to each city with demand greater than the mammography unit's capacity. Furthermore, no modification of this allocation is made by any procedure (local search or shaking) during the search.

Subsection 5.1 describes the proposed VNS-based algorithm, and the following subsections describe its modules.

5.1   Variable Neighborhood Search

The Variable Neighborhood Search (VNS) algorithm [9] developed for the MULP is a basic VNS [10] and works according to Algorithm 1.

Algorithm 1 : VNS

1: $s_0 \leftarrow InitialSolution()$
2: $s \leftarrow LocalSearch(s_0)$
3: **while** Stopping criterion is not satisfied **do**
4:      $k \leftarrow 2$;
5:      **while** $k \leq r$ **do**
6:          $s' \leftarrow Shaking(s, k)$
7:          $s'' \leftarrow LocalSearch(s')$
8:          **if** ($f(s'') > f(s)$) **then**
9:              $s \leftarrow s''$
10:            $k \leftarrow 2$
11:        **else**
12:            $k \leftarrow k + 1$
13:        **end if**
14:    **end while**
15: **end while**
16: Return s;

Algorithm 1 begins in line 1 by constructing an initial solution according to Subsect. 5.4. Next, in line 2 it is refined by the local search procedure described in Subsect. 5.6. In order to avoid getting stuck in a local optimum, the algorithm goes into a loop that works as follows. Initially, all mammography units of k cities are removed, as well as all links associated with these k cities. In the next step, the solution is restored by a constructive mechanism described in Subsect. 5.4. Both steps (removal and construction) compose the Shaking procedure, which is described in Subsect. 5.7. In line 7, the current solution is refined. If this local optimum solution s'' is better than the current solution s (line 8), then s is updated and the level of perturbation returns to its minimum value ($k = 2$); otherwise, the perturbation is increased (line 12). The algorithm ends when the stopping criterion is satisfied.

5.2   Solution Representation

A solution s of the MULP is represented by a tuple $s = (u, v)$, where u and v are vectors, both of size n. Each position j of the vector u shows that the city j is covered by some equipment installed in the city u_j. If u_j assumes the value 0, it means that city j is not covered by any mammography unit. Each index j of the vector v indicates that the city j holds a total of v_j mammography units.

An example of a solution to the problem is shown in Table 2. In this case, $p = 2$ mammography units are available to cover up to $n = 8$ cities. The first line of the table shows the indexes of the cities; the second line, u_j, shows the links between cities, for example: cities 1, 2, and 3 are all covered by city 1, while cities 4 and 5 are covered by city 5. Cities 6, 7, and 8 are not covered by any city. Finally, the last line corresponds to the vector v that stores the number of equipment per city; in the case, one mammography unit was allocated to city 1 and another to city 5.

In Fig. 1(a), a map is displayed with the spatial distribution of 8 cities and 2 mammography units to be located. Each vertex in the map corresponds to a city and the number positioned inside the circle indicates the index of the respective city. A vertex in red color indicates that the respective city receives equipment. Thus, the mammography units were installed in cities 1 and 5. An edge connects the cities that will be served by these equipment. For example, edges $(1, 2)$ and $(1, 3)$ show that cities 2 and 3 are covered by city 1.

Table 2. Solution example

j	1	2	3	4	5	6	7	8
u_j	1	1	1	5	5	0	0	0
v_j	1	0	0	0	1	0	0	0

5.3 Evaluate Function

A solution $s = (u, v)$ is evaluated by the function f, given by Eq. (12), which must be maximized:

$$f(s) = \sum_{j \in N \mid u_j \neq 0} dem_j \qquad (12)$$

where N is the set of cities, dem_j represents the demand for mammography screenings of the city j, and u_j is a variable that assumes a non-zero value if the city j is covered by some city and value 0, otherwise. The objective is to maximize the total number of screenings using p mammography units.

5.4 Initial Solution

We present below a constructive heuristic procedure to generate a solution for the MULP.

Step 1: Calculate, for each city i yet not covered, the demand of its coverage region, including its own demand for mammography screenings;

Step 2: Sort the cities, in decreasing order according to the demand from each region covered;

Step 3: Calculate the number of mammography units that are necessary to cover the city i that has the greatest demand. If this number is less than or equal to the number of mammography units available, cover the total demand of this city and update the available number of mammography units; otherwise, return to Step 1. In both cases, remove this city from the list;

Step 4: Calculate the amount of idle mammography screenings of the equipment installed in the city i and determine the cities in the region that are not covered;

Step 5: Solve the knapsack problem, considering the city i as a knapsack of capacity equal to the amount of idle mammography screenings and as items, the demands for mammography screenings of the cities still not covered that belong to its region;

Step 6: Assign the cities returned by Step 5 to the city i;

Step 7: If the remaining number of mammography units is still greater than zero, return to Step 1; otherwise, finalize the procedure and return the cities in which the mammography units will be installed, as well as the cities covered by them.

Note that the cities in this constructive procedure are initially sorted in decreasing order of the total number of mammography screenings demanded by each city of the region that it can cover, i.e., the demand of the city itself is summed up with the demands of the cities that are in its coverage region. The coverage region of a city i is composed by all cities j which are at distance $d_{ij} \leq R$ km and $d_{ji} \leq R$ km from it.

In this sorting, only the cities that have hospital infrastructure are candidates, here considered those that have a high demand for mammography screenings, that is, greater than $demMin$ ones. This value is adopted to economically justify the installation of mammography units in a city. Whenever a mammography unit is allocated to a city i, it is considered that this city will be fully covered. The amount of idle mammography screenings of this equipment is used to cover the demand of the cities that are in its coverage region.

The choice of which cities will be covered by the city i is done by solving the 0–1 Knapsack Problem (KP). The KP consists of filling a knapsack of capacity W with items of different weights and profits. The goal is to fill the knapsack with the highest possible profit so that it does not exceed its capacity. The following analogy is made in the construction of the initial solution for this problem: the uncovered cities of the region correspond to the items that can be inserted in the knapsack; each city or item has a profit and a weight and both match the demand for screenings from the respective city. Finally, the knapsack is represented by the city i, which has a mammography unit with idle capacity $W = cap - w_i$, where cap is the capacity of the equipment and w_i is the demand for mammography screenings of the city i. The method returns the list of cities that will be covered by the mammography unit installed in the city i.

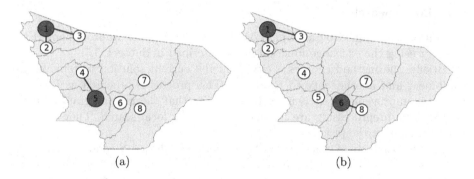

(a)   (b)

Fig. 1. Illustration of the exchange move (Color figure online)

We apply a dynamic programming algorithm[1] to solve the knapsack problem. Thus, the solution returned by it is exact. It has pseudo-polynomial time complexity, that is, $\mathcal{O}(mW)$, where m is the amount of items and W is the capacity of the knapsack. For the largest instance ($m = 310$ and $W < cap = 5069$), the problem is solved instantly.

5.5 Neighborhood Structure

The exploration of the solution space is made using a single type of move, called exchange function. This move consists of removing a mammography unit from a city and then allocating it to another city that has infrastructure to receive it but does not yet have that equipment.

---

[1] Its code was extracted from https://www.geeksforgeeks.org/0-1-knapsack-problem-dp-10/.

An application example of this move is shown in Figs. 1(a) and (b). In these figures, each vertex corresponds to a city, and the ones highlighted in red are the cities that have mammography units. The edges connecting two vertices indicate service dependency. In Fig. 1(a), for example, it is possible to verify that cities 1 and 5 have equipment and they cover the cities $\{1, 2, 3\}$ and $\{4, 5\}$, respectively.

In these figures, it can be seen that one mammography unit was removed from the city 5 and thus the city 4 is no longer covered by the city 5. Then it is necessary to choose another city that respects the restrictions to receive the equipment that is now available. In this example, city 6 was chosen to illustrate this move, as we can see in Fig. 1(b). Once these choices have been made, the next step is to determine which cities will be covered by city 5. Thus, the knapsack problem is solved. After applying the dynamic programming procedure mentioned before, city 8 was chosen to be served by city 6. Finally, in Fig. 1(b), we have the final solution resulting from the application of the exchange function move. The mammography unit of the city 5 is now in city 6 and this, in turn, only covers itself and the city 8.

5.6 Local Search

A solution s is refined by an Uphill Method with the First Improvement (FI) strategy using the exchange function move described in Subsect. 5.5.

Briefly, the method chooses a city from the current solution to remove its equipment and another city to receive it. This process is repeated as long as there are improvements. It is important to note that the mammography units allocated by preprocessing are not modified by the local search. They remain fixed throughout the procedure.

Initially, two sets of cities, S and V, are formed. The set S contains all cities that can host mammography units but do not yet have them, and the set V contains all cities that have at least one equipment. These sets are scrambled at each procedure call to prevent the method from always considering the same choice of cities in different runs.

Then, for each city $i \in V$, we remove its equipment. The next step is to choose a city $j \in S$ to receive such equipment. To fulfill this task, for each city $j \in S$ we solve a knapsack problem in order to choose the cities of the coverage region of the city j to be served. If the exchange move improves the current solution, it is accepted and the method is reset; otherwise, the method proceeds to the next city $j \in S$.

The method ends when all possible exchanges do not generate an improvement solution, thus ensuring that the returned solution is a local optimum in relation to the neighborhood used.

The uphill method described in Algorithm 2 takes as input a solution s. Initially it builds the sets S and V based on s (lines 5 and 6). In line 10, the RemoveEquipment(i) method is called. It is responsible for removing the mammography unit from city i and eliminating its dependencies, thus returning an incomplete solution s. In line 11, the loop that iterates on the cities that have infrastructure to receive an equipment starts. In line 12, the InsertEquipment(j)

method is responsible for inserting an equipment in the city j, returning a new solution s'. Then, the knapsack problem is performed in order to define which cities in the coverage region of city j will be covered. If the solution s'' returned by the knapsack improves the best solution found so far, then it is updated and the method restarts from line 3. Otherwise, the loop continues with the next city j. Finally, if the loop in line 3 ends without improvement, we terminate the method and return the solution s_{best} found with the assurance that we find a local optimum for this neighborhood.

5.7 Shaking Procedure

The shaking procedure works as follows. All mammography units and their dependencies of k cities ($k \leq r$) are removed from the current solution s (except those related to the preprocessing phase) and then this solution is restored by solving the knapsack problem according to Subsect. 5.4.

After the removal operation, the cities that will host the mammography units are selected in a partially greedy way as follows. For each city i that has not a mammography unit and that can host it, we calculate its potential of service, that is, the sum of the demands of mammography screenings of the coverage region of that city i. In other words, the service potential is calculated by summing the

---

Algorithm 2: First Improvement

Require: Solution s

  1: $s_{best} \leftarrow s$
  2: $hasImprovement \leftarrow true$
  3: **while** $hasImprovement$ **do**
  4:     $hasImprovement \leftarrow false$
  5:     $S \leftarrow \{$Available cities in solution s that can host equipment$\}$
  6:     $V \leftarrow \{$Cities in solution s that have equipment$\}$
  7:     Shuffle(S)
  8:     Shuffle(V)
  9:     **for** $i \in V$ **do**
10:         $s \leftarrow$ RemoveEquipment(i)
11:         **for** $j \in S$ **do**
12:            $s' \leftarrow$ InsertEquipment(j, s)
13:            $s'' \leftarrow$ KnapsackProblem(j, s')
14:            **if** $f(s'') > f(s_{best})$ **then**
15:               $s \leftarrow s''$
16:               $s_{best} \leftarrow s''$
17:               $hasImprovement \leftarrow true$
18:               Goto line 3
19:            **end if**
20:         **end for**
21:     **end for**
22: **end while**
23: return s_{best}

demand of all cities that are less than 60 km from city i. The partially greedy choice is made so that different solutions are analyzed. We chose one of the four cities that have the greatest service potential.

Then, for each city i chosen to host an equipment, it is also necessary to define its service dependencies, that is, which cities in its coverage region will be served by the city i. For this decision we solve the knapsack problem (Subsect. 5.4), obtaining, thus, the greatest possible demand that the city i can cover.

After that, the city chosen to host the equipment and the cities it serves are included in the current solution, and the solution restoration method proceeds by recalculating the service potential of the remaining cities as previously presented.

6 Computational Experiments

The mathematical programming model presented in Sect. 4 was implemented in the Gurobi solver, academic version 8.0.0, with default settings, while the proposed VNS algorithm, presented in Sect. 5.1, was developed in C++ language. To test them was used an Intel Core i5 @ 2.5 GHz computer, with 8 GB of RAM under the Ubuntu 18.04 operating system.

Table 3. Characteristics of the instances

Instance	# Cities	# Cities with infrastructure	# Total equipment	# Preproc. equipment	# Remaining equipment
1	853	420	310	114	196
2	853	420	261	114	147
3	853	420	212	114	98
4	853	420	163	114	49
5	142	73	55	19	36
6	142	73	46	19	27
7	142	73	37	19	18
8	142	73	28	19	9

For testing the methods, 8 instances were used. These instances refer to female population data for the year 2010 of the Minas Gerais State, Brazil, and they are available at http://www.decom.ufop.br/prof/marcone/projects/ MULP/instances-MG-2010.rar. This State has 853 cities and according to the sector of statistics of its State Secretary for Health (SES/MG), there were 310 mammography units in July of 2018 and the total demand was 1293968 mammography screenings. Considering the current equipment location and the problem characteristics described in Sect. 4, it is possible to perform 970103 mammography screenings, that is, only 75% of total demand. The distances between cities in instances 1 to 4 were obtained by applying the formula of Euclidean distance between cities, while in instances 5 to 8 these values refer to real distances obtained through Google Maps with travel by car. We consider that a city has hospital infrastructure if it has the demand for at least $demMin$ mammography screenings. In our case, we set $demMin = 500$. The maximum travel distance to be served was set at $R = 60$ km, a value that is recommended by the Health Ministry of Brazil. The scenarios differ by the number of mammography units available for allocating and the number of cities considered. In the first four instances, the whole State is considered, while in the last four ones only cities that are 100 km from Ouro Preto city, except Belo Horizonte, are considered.

Table 3 summarizes the characteristics of these instances. Column 1 shows the instance number. The second column indicates the total number of cities and the third one shows the number of cities that have the infrastructure to receive equipment. The fourth column reports the number of equipment in the instance. The fifth column shows the number of equipment used by the preprocessing strategy established at the beginning of Sect. 5. Finally, the last column reports the amount of equipment available for applying the methods.

For calibrating the parameters of the VNS algorithm, we test empirically the following values: $r \in \{4, 6, 8, 10\}$ as the number of mammography units removed in the Shaking procedure according to Subsect. 5.7 and $iterMax \in \{100, 200, 300, 400\}$ as the maximum number of iterations without improvement. The best values found empirically were $r = 8$ and $iterMax = 100$.

Table 4. Results Gurobi × VNS

Inst.	Demand preproc.	Gurobi			VNS		
		ub	Best	Time (s)	Best	Average	Time (s)
1	577866	1291621	**1291621**	3.37	**1291621**	1291621	1.25
2	577866	1290753	**1288076**	3600.00	1278820	1276812	2466.89
3	577866	1074628	1074466	3600.00	**1074563**	1074541	3600.00
4	577866	826247	826225	3600.00	**826247**	**826247**	19.80
5	96311	221140	**221140**	0.94	**221140**	220690	3.32
6	96311	221140	**221140**	102.17	**221140**	216462	71.53
7	96311	187553	**187544**	3600.00	187478	187386	326.37
8	96311	141932	**141932**	2.95	**141932**	141930	494.65

Table 4 reports the results considering that a mammography unit performs 5069 mammography screenings per year [14]. Column 1 shows the instance, and column 2 indicates the demand served by preprocessing (114 equipment were allocated after this phase for the instances 1 to 4 and 19 for the instances 5 to 8). Columns 3–5 show the upper bound and the value returned by the Gurobi solver with the remaining equipment, as well as the time consumed for solving the instance. Since the solver has been applied only for 3600 s of processing time, then this value returned by Gurobi is optimal if the time spent by it is less than 3600 s. The last three columns show the best result, the average result and the time spent by the VNS algorithm, respectively. The columns "Best" and

"Average" represent the total demand, including the preprocessing one. Values highlighted in bold indicate the best solution for the instance.

According to Table 4, in instance 1, which represents the real instance of Minas Gerais State, both the VNS algorithm and Gurobi Solver are able to find the optimal solution quickly. The total demand met, of 1291621 mammography screenings, is higher than the current allocation of 970103 ones. We can also observe that the VNS algorithm achieves the optimal solution in 5 instances and produces a better solution than Gurobi in two instances (instances 3 and 4). Only in two instances (2 and 7) the VNS algorithm did not overcome Gurobi.

Figure 2 illustrates a typical evolution of the best solution' value produced by the VNS algorithm over the time. As we can see, the VNS method is able to improve the value of the solution over the time.

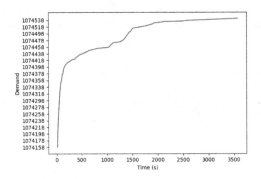

Fig. 2. Evolution of the best solution: instance 3

7   Conclusions and Future Work

This work addressed the Mammography Unit Location Problem (MULP). To solve it, a mathematical programming formulation and a VNS-based heuristic algorithm were developed. In order to test them, eight instances related to data from the State of Minas Gerais, Brazil, were used.

The proposed algorithm was able to produce good quality solutions and outperform the Gurobi solver in two instances. The variability of the final solutions is also low, except in instance 6. Besides it, the algorithm was able to improve the value of the solution over time. It is interesting to note that the total demand of the Minas Gerais State was almost fully covered with the existing mammography units. In fact, the demand served was 99.8% in instance 1. On the other hand, the demand covered by the two models (exact and heuristic) was much higher than the present one. This is due to the fact that the allocations of these equipment have much political influence.

The variability of the final solutions of the VNS algorithm can be reduced by adequately calibrating its parameters. This can be done, for example, by using the Irace package [15]. In addition, other neighborhood structures can be designed to improve its performance.

We also suggest as future work to consider: (1) more recent data of the female population in the age range indicated for mammography screenings; (2) the current grouping of cities in health regions; (3) the real distances between cities; (4) that a city can be partially covered by an equipment and (5) the proposition of itineraries for mobile mammography units to cover cities not covered by the current location of the mammography units.

Acknowledgments. The authors thank the Brazilian agencies FAPEMIG (grant PPM-CEX 676/17), CNPq (grants 438473/2018-3, 428817/2018-1 and 307915/2016-6), CAPES and the Federal University of Ouro Preto for supporting this study.

References

1. Aday, L.A., Andersen, R.: A framework for the study of access to medical care. Health Serv. Res. **9**(3), 208–220 (1974)
2. Ahmadi-Javid, A., Seyedi, P., Syam, S.S.: A survey of healthcare facility location. Comput. Oper. Res. **79**, 223–263 (2017)
3. Amaral, P., Luz, L., Cardoso, F., Freitas, R.: Spatial distribution of mammography equipment in Brazil. Revista Brasileira de Estudos Urbanos e Regionais **19**(2), 326–341 (2017). (in Portuguese)
4. Andrade, M.V., et al.: Spatial distribution of mammography equipment in Minas Gerais and the effect on probability of doing the mammography screening. In: Proceedings of the XX Encontro Nacional de Estudos Populacionais, pp. 1–21. ABEP, Foz do Iguaçu (2016). (in Portuguese). http://twixar.me/20zT. Accessed 10 Sept 2018
5. Church, R., Velle, C.R.: The maximal covering location problem. Pap. Reg. Sci. Assoc. **32**, 101–118 (1974)
6. Corrêa, V.H.V., Lima, B.J.C., Silva-e-Souza, P.H., Penna, P.H.V., Souza, M.J.F.: Mammography unit location problem: a case study in the public health care. In: Proceedings of the L Brazilian Symposium of Operations Research, Rio de Janeiro, Brazil (2018). (in Portuguese). http://twixar.me/60zT. Accessed 01 Dec 2018
7. Garey, M.R., Johnson, D.S.: A Guide to the Theory of NP-Completeness. WH Freemann, New York (1979)
8. Hamer, L.: Improving Patient Access to Health Services: A National Review and Case Studies of Current Approaches. Health Development Agency, London (2004)
9. Mladenović, N., Hansen, P.: Variable neighborhood search. Comput. Oper. Res. **24**, 1097–1100 (1997)
10. Hansen, P., Mladenović, N., Todosijević, R., Hanafi, S.: Variable neighborhood search: basics and variants. EURO J. Comput. Optim. **5**(3), 423–454 (2016). https://doi.org/10.1007/s13675-016-0075-x
11. INCA: Parameters for the screening of breast cancer: recommendations for state and municipal managers. INCA (2009). (in Portuguese). http://twixar.me/M0zT. Accessed 01 Sept 2018
12. INCA: Guidelines for the early detection of breast cancer in Brazil. INCA (2015). (in Portuguese). http://www.saude.pr.gov.br/arquivos/File/Deteccao_precoce_CANCER_MAMA_INCA.pdf. Accessed 04 Sept 2018

13. INCA: Inca estimates that there will be 596,070 new cases of cancer in 2016. INCA (2015). (in Portuguese). http://twixar.me/c0zT. Accessed 03 Mar 2019

14. INCA: Review of the parameter for calculation of the production capacity of the simple mammography unit. INCA (2015). (in Portuguese) http://twixar.me/N0zT. Accessed 10 Sept 2018

15. López-Ibáñez, M., Dubois-Lacoste, J., Cáceres, L.P., Birattari, M., Stützle, T.: The irace package: iterated racing for automatic algorithm configuration. Oper. Res. Perspect. **3**, 43–58 (2016)

16. Sathler, T.M., Conceição, S.V., Almeida, J.F., Pinto, L.R., de Campos, F.C.C., Miranda Júnior, G.: Location and allocation problem of specialized medical centers in Minas Gerais State. In: Proceedings of the XLIX Brazilian Symposium of Operations Research, Blumenau, Brazil, pp. 2988–2999 (2017). (in Portuguese)

17. Torre, L.A., Islami, F., Siegel, R.L., Ward, E.M., Jemal, A.: Global cancer in women: burden and trends. Cancer Epidemiol. Prev. Biomark. **26**(4), 444–457 (2017). https://doi.org/10.1158/1055-9965.EPI-16-0858

18. Villar, V.C.F.L., de Souza, C.T.V., Delamarque, E.V., de Seta, M.H.: Distribution of mammography units and mammography exams in the Rio de Janeiro State in 2012 and 2013. Epidemiologia e Serviços de Saúde **24**, 105–114 (2015). (in Portuguese)

19. Xavier, D.R., de Oliveira, R.A.D., de Matos, V.P., Viacava, F., de Campos Carvalho, C.: Coverage of mammography units, allocation and use of equipment in health regions. Saúde em debate **40**(110), 20–35 (2016). (in Portuguese)

A Hybrid Heuristic Algorithm
for the Dial-a-Ride Problem

André Luyde S. Souza[1]([✉]), Jonatas B. C. Chagas[1,2], Puca H. V. Penna[1],
and Marcone J. F. Souza[1]

[1] Departamento de Computação, Universidade Federal de Ouro Preto,
Ouro Preto, Brazil
andreluyde@iceb.ufop.br, {puca,marcone}@ufop.edu.br
[2] Departamento de Informática, Universidade Federal de Viçosa, Viçosa, Brazil
jonatas.chagas@ufv.br

Abstract. In this paper, we propose a simple heuristic algorithm based
on the Variable Neighborhood Search (VNS), which combines with the
Set Covering strategy in order to solve the Dial-a-Ride Problem (DARP).
In this problem, customers must be served by a heterogeneous fleet of
vehicles. Each customer has a pickup and a delivery location, where
each one of them has time windows that must be obeyed. All vehicles
have a duration time and have to start and end their routes in a single
depot, and each customer has a maximum time ride. We have tested our
algorithm on the benchmark instances of literature. Experiments showed
that although the algorithm is simple, it can obtain the optimal solutions
for some instances and achieve solutions near the optima for the others.

Keywords: Dial-a-ride · Vehicle routing · Pickup and delivery ·
Variable Neighborhood Search · Randomized Variable Neighborhood
Descent · Metaheuristic

1 Introduction

Lutz et al. [14] have reported the fact of the combinations of declining fertility,
and increasing life expectancies resulted in the aging of the population. They
also presented a likely increase in the speed of the aging population in the next
decades and continuous aging of the world's population throughout the century.
Due to this aging of the population, a lot of services need to be improved. Among
them are transportation services. The problems involving the transportation
can be classified as Vehicle Routing Problems (VRP). The VRP consists of
determining a set of routes executed by a fleet of vehicles to satisfy the necessities
of a set of users [22]. Many VRPs with different characteristics have been studied
over the years, and one of them is the Dial-a-Ride Problem - DARP [4].

The DARP arises in the context of users transportation, a service for the
people with reduced locomotion, or disabled people. Different from other VRPs,

Supported by CAPES.

R. Benmansour et al. (Eds.): ICVNS 2019, LNCS 12010, pp. 53–66, 2020.
https://doi.org/10.1007/978-3-030-44932-2_4

in the DARP, the customer's inconvenience has to take into account. The customer's nuisance can be defined as waiting time and travel time. The DARP consists of planning routes and schedules to meet a set of customers from a fleet of vehicles. Each customer has a pickup and a delivery location, where it must be collected and delivered, respectively. A maximum ride time has to be respected for each customer and duration time for each vehicle. A time window must be respected for every pickup and delivery point. The aim is to design a routing plan, i.e., a set of routes, capable of attending as many users as possible, under a set of constraints. Not different from other VRP variants, the DARP can be classified as static and dynamic. In the static case, all information about transportation request is known beforehand and used to calculate an a priori routing plan. In the dynamic case, the routing plan is designed in a real-time manner when a new request is revealed [6].

The DARP can be classified as the combination of two classical optimization problems: Vehicle Routing Problem with Pickup and Delivery (VRPPD) and Vehicle Routing Problem with Time Windows (VRPTW). The DARP differs from them as far as the human perspective is concerned, in DARP the comfort and convenience of users should be taken into account [6].

Besides, DARP is classified as an $\mathcal{NP}$-Hard problem due to its high complexity. Thus, optimal solutions cannot be obtained in a reasonable computational time. Therefore, DARP has been mainly addressed by heuristic and metaheuristic algorithms [6,10].

In this paper, we propose a VNS-based algorithm to solve the DARP. Our algorithm was combined with a Set Covering Problem (SCP) formulation, which is applied throughout the VNS algorithm in order to improve its current solution. This approach has been able to find high-quality solutions in short computational time.

The remainder of this paper is organized as follows. In Sect. 2, we present a brief literary review, contextualizing and exemplifying the DARP and its variants, as well as different ways for solving them. The formal definition is shown in Sect. 3. The proposed algorithm is detailed in Sect. 4. Section 5 presents the computational experiments and results. Conclusions are given in Sect. 6.

2   Literature Review

Several DARP variants have been studied in the past few years. We can classify them into two main branches: static and dynamic versions.

In the static version, all the information about the customer requests and vehicles is known beforehand. Cordeau and Laporte [4] considered the problem with a homogeneous fleet of vehicles and single depot. To solve it, they used a Tabu Search (TS) algorithm combined with three heuristic methods. Cordeau [5] approached the DARP by an exact branch-and-cut algorithm, which was tested on a set of instances randomly generated with a maximum of 48 requests. Jorgensen et al. [12] addressed the variant of DARP with a Genetic Algorithm (GA), which considers a heterogeneous fleet of vehicles and multiple depots they made the algorithm basing on the classical cluster-first and route-second approach. Mauri et al. [17], using the data provided by the city of Vitória–ES in Brazil, approached a static variant with a heterogeneous fleet of vehicles and multiple depots. The authors used a Simulated Annealing (SA) algorithm to solve the problem. Their results show that SA produced good quality solutions for all the instances.

Parragh [19] combined other constraints to the classic DARP and proposed a branch-and-cut algorithm to solve the problem. She added real characteristics to the problem as heterogeneity to the vehicles and users. In her tests, for the instances up to 40 requests, she found the optimal solution. Again, Parragh et al. [20] proposed a hybrid column generation and large neighborhood search algorithm to solve the problem. They improved nine of the 20 instances used in their tests in 1.1%. Currently, the best-known solutions of eight of these instances are the solutions founded by Parraagh et al. [20].

Braekers et al. [3] approached the DARP using the branch-and-cut algorithm combined with a metaheuristic named determinist annealing based in the simulated annealing metaheuristic. They formulated the problem as a single depot and multi-depot for heterogeneous DARP. They tested their approach in benchmarks created by Cordeau [5], Parragh 2011 [19], and Cordeau et al. [4]. They also extended the sets of instances creating 36 larger new instances in the same way of Parragh [19]. They match or exceed most of the best-known results found for these instances. Most of the solutions founded by the authors remain as the best-known solution for the instances. Masmoudi et al. [16] solved the same heterogeneous version of the DARP using a hybrid Genetic Algorithm. They used a constructive heuristic and efficient crossovers to guided the algorithm. They tested their approach in the instances of [5,19], and [3]. The results demonstrate the algorithm is efficient in terms of best and average solution quality.

In the dynamic version of DARP, new customer requests are added during the route. Madsen et al. [15] solve the DARP adapting an insertion algorithm called REBUS originally developed by Jaw et al. [11]. The version of the problem solved used the multiple capacities and multiple objectives. Using data given by Copenhagen Fire-Fighting Service, they tested their approach on real-life cases. In a flexible way, the algorithm permits a weighing of multiple goals such that the solution reflects the customer's preferences. Attanasio et al. [1] approached the dynamic DARP with the objective to accept as many requests as possible while satisfying the constraints of the problem. To solve the problem, they used a parallel strategy of a Tabu Search heuristic previously used by Cordeau and Laporte [4] in the static case of the problem.

In Germany, Beaudry et al. [2] proposed a two-phase heuristic to solve a dynamic DARP. The first phase is a simple insertion scheme, and the second phase is a Tabu Search algorithm that considers infeasible solutions during the search process. The problem was structured with data given by several large hospitals. They proposed additional constraints on the standard problem. For example, a different degree of urgency in the requests, the patient of the request cannot share the vehicle with another patient, among others. Experiments provided high-quality solutions, and the algorithm show capable of handling the dynamic aspect of the problem. Schilde et al. [21] proposed a dynamic DARP to solve the problem of the organization performing patient transportation in Austria. They approached the dynamic case of the DARP with a homogeneous fleet of vehicles and a single depot. To solve the problem, the authors proposed four different metaheuristics, a Variable Neighborhood Search (VNS), a Stochastic Variable Neighborhood Search (S-VNS), both dynamics, a Multiple Plan Approach (MPA), and a Multiple Scenario Approach (MSA). They tested their algorithms on a set of 12 instances based on a real road network. Results show that dynamic S-VNS strongly outperforms the others. Again in Germany, Hanne et al. [8] solve a dynamic DARP with hospital-specific constraints. They designed a computer-based planning system, Opti-TRANS, to support several concerns related to patient transport. They illustrated the system work in the daily performance of a large German hospital, and the system presents many benefits, including streamlined transportation processes and workflow.

Thus, the large number of DARP variants, here, we study the Heterogeneous DARP with single depot proposed by Parragh et al [20]. The heterogeneity is considered in the user and vehicles [20].

3   Formal Definition

The DARP can be defined as follows. Let $G = (V, E)$ be a complete graph, where V is the set of vertices and E the set of arcs. The set of vertice V is partitioned into $\{\{0, 2n + 1\}, P, D\}$, where 0 and $2n + 1$ are two copies of the depot, $P = \{1, ..., n\}$ is the set of pickup vertices and $D = \{n + 1, ..., 2n\}$ is the set of delivery vertices. For each arc $(i, j) \in E$ is associated a routing cost c_{ij} and a travel time t_{ij}.

Each customer request i consists of a pickup and delivery vertex pair $\{i, n+i\}$, where $i \in P$ and $n + i \in D$. The maximum travel time of each customer cannot exceed L. To each vertex $i \in V$ is associated a load q_i, with $q_0 = q_{2n+1} = 0$, $q_i \geq 0 \ \forall i \in P$ and $q_i = -q_{i+n} \ \forall i \in D$, and a non-negative service time τ_i. Moreover, each $i \in V$ has a time window $[e_i, l_i]$, where e_i and l_i are integers non-negative.

The set of vehicles is represented by K. Each vehicle $k \in K$ starts and ends its route in the depot. The capacity of vehicle k is Q_k and the maximal duration of route k is denoted by T_k.

Given the vertex v_i we denoted by A_i the arrival time of the vehicle in the vertex; and B_i the beginning of the service, where $B_i \geq \max\{A_i, e_i\}$, cannot start before e_i. The departure time $D_i = B_i + \tau_i$ is the time the vehicle leaves the vertex. The vehicle waiting time is defined by $W_i = B_i - A_i$. The ride time of the client is determined by $L_i = B_{n+i} - D_i$ and the total duration of the route is calculated as $B_{n+1}^k - B_0^k$, where B_{n+1}^k and B_0^k represents the beginning of service on the depot by vehicle k, when it finishes and starts the ride, respectively.

The objective of the DARP is to find a set of routes that serve all customers such that minimizes the total routing cost.

4 Heuristic Approach

To solve the problem we based our heuristic on a VNS and a set covering procedure. Sections 4.1–4.4 define the heuristic components, while Sect. 4.5 presents the set covering model.

4.1 Solution Representation

We represent a DARP solution through a matrix of $|K|$ rows, where each row informs the route assigned to a vehicle (the k-th row refers to the route of the k-th vehicle). Each route begins and ends at the depot. The pickup and delivery customers are inserted in the route, satisfying the problem constraints.

Figure 1 shows a solution representation for an instance with five vehicles $(A, B, \cdots, E)$ and 13 customer requests. The numbers with positive sign $(1+ \cdots 13+)$ represent the pickup locations and the ones with negative sign $(1- \cdots 13-)$ are the delivery points. The depot is represented by node 0.

A	0	9+	4+	9-	2+	2-	4-	6+	6-	0
B	0	3+	5+	3-	1+	1-	5-	0		
C	0	7+	8+	7-	8-	0				
D	0	10+	12+	13+	10-	12-	13-	0		
E	0	11+	11-	0						

Fig. 1. Example of a solution representation.

4.2 Solution Evaluation

The solution evaluation was done based on an approach used by Cordeau and Laporte [4] and Parragh et al. [18]. Following them, we penalized the violations of load $q(s)$, duration $d(s)$, time windows tw, and user ride time $t(s)$. The load and duration are computed and penalized in the route, based on the constraints Q_k and T_k. The time windows penalization is computed as

$$tw = \sum_{i=0}^{2n} (B_i - l_i)^+$$

where $x^+ = \max\{x, 0\}$. In turn, the ride time is calculated as

$$t = \sum_{i=1}^{n} (L_i - L)^+$$

Thus, the solution evaluation was calculated as follows:

$$f(s) = c(s) + \alpha tw(s) + \beta t(s) + \delta d(s) + \gamma q(s)$$

The penalization variables for load (γ), duration (δ), time window (α), and ride time (β) violations were set to $\alpha = \beta = \delta = \gamma = 1$.

4.3   Neighborhood Structures

To explore the solution space of the DARP, six neighborhoods was implemented. These neighborhoods consist in the relocation of requests among the solution routes. For all neighborhoods, two routes, r_1 (blue line) and r_2 (black line) are randomly selected.

– Relocation – A request (pickup and delivery points) are removed from r_1 and inserted in r_2. Figure 2 shows the relocation of request $(1^+, 1^-)$ from r_1 to r_2.

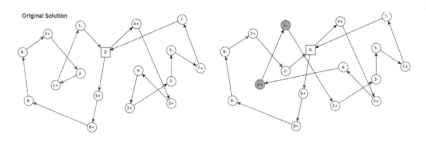

Fig. 2. Relocation. (Color figure online)

– Swap – Two requests are selected, one from r_1 and another and r_2, and exchanged them. Figure 3 shows the exchange of one request $(6^+, 6^-)$ in route r_1 with other request $(4^+, 4^-)$ in route r_2.

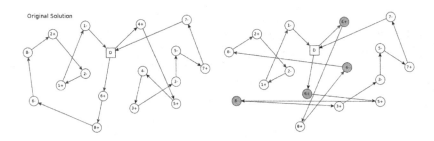

Fig. 3. Swap. (Color figure online)

– Crossover – Two points are selected, one in r_1 and other in r_2. All pickup customers before the selected point in r_1 (and their respective deliveries) are connected to all pickup customers (and their respective deliveries) that come after the selected point in r_2. The same way to the customers before r_2 point and after the r_1 point. Figure 4 shows the crossover of routes, the red line shows the selected points for each route and the relocation of requests.

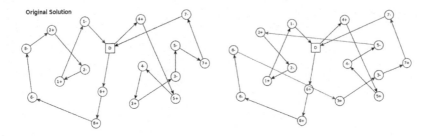

Fig. 4. Crossover. (Color figure online)

- Swap(2) – Four requests are selected, two adjacent requests from r_1 and two adjacent requests in r_2, and they are exchanged. All the four possible combination orders of exchanging the request arcs are considered. Figure 5 shows the exchange of two requests $(6^+, 6^-)$ and $(8^+, 8^-)$ in route r_1 with two requests $(4^+, 4^-)$ and $(3^+, 3^-)$ in route r_2.

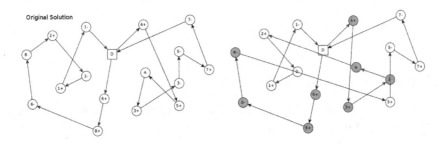

Fig. 5. Swap2. (Color figure online)

- Swap(2-1) – Three requests are selected, two adjacent requests from r_1 and one in r_2, and they are exchanged. The move examines the two possible visiting orders of transferring the r_1 requests. Figure 6 illustrates the exchange of two requests $(6^+, 6^-)$ and $(8^+, 8^-)$ in route r_1 with the request $(4^+, 4^-)$ in route r_2.
- Relocation(2) – Two adjacent requests are removed from r_1 and inserted in r_2. Figure 7 shows the relocation of requests $(6^+, 6^-)$ and $(8^+, 8^-)$ from r_1 to r_2.

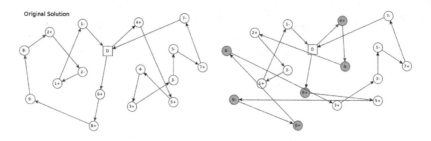

Fig. 6. Swap2. (Color figure online)

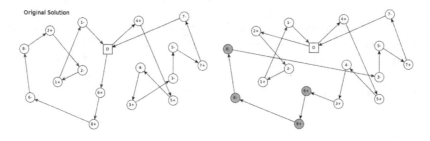

Fig. 7. Relocation. (Color figure online)

4.4   Variable Neighborhood Search

To tackle the DARP, a VNS-based algorithm [9] was proposed. For its use, three components have to be specified:

- Generate initial solution Procedure: A function that greedily selects each request taking into account the time window allocating to each chosen car with the shortest time traveled.
- Local Search Procedure: The method used as local search was the Randomized Variable Neighborhood Descent (RVND), that has its behavior like a classic Variable Neighborhood Descent, but the neighborhoods are randomly selected.
- Shaking Procedure: Let Ω be a set of neighborhood structures. The shaking procedure consists of selecting a random neighborhood belonging to Ω and then choosing a random neighbor of the current solution in this neighborhood.

All steps of our VNS algorithm applied to solve the DARP are presented in Algorithm 1. The algorithm receives as parameters an initial solution s_0 and the maximum number of iterations (*iterMax*). In the line 4 is applied a shaking in the current solution. The local search is applied in line 5, where all different feasible routes are stored in a set R. This set is used in the set covering model. When the number of iterations exceeds half of *iterMax*, the set covering procedure is applied (line 13).

Algorithm 1. VNS procedure

1: **procedure** VNS (s_0, $iterMax$)
2:    $iter \leftarrow 0$
3:    **while** $iter \leq iterMax$ **do**
4:       $s' \leftarrow$ Shaking(s)
5:       $s'' \leftarrow$ LocalSearch(R, s')
6:       **if** $f(s'') < f(s)$ **then**
7:          $s \leftarrow s''$
8:          $iter \leftarrow 0$
9:       **else**
10:          $iter \leftarrow iter + 1$
11:       **end if**
12:       **if** $iter \geq iterMax/2$ **then**
13:          $s'' \leftarrow$ setCovering(R)
14:          **if** $f(s'') < f(s)$ **then**
15:             $s \leftarrow s''$
16:             $iter \leftarrow 0$
17:          **end if**
18:       **end if**
19:    **end while**
20:    **return** s
21: **end procedure**

4.5 Set Covering

Let R be a set of different feasible routes found by VNS algorithm, V the set of customers and $|K|$ the maximum number of vehicles available. A cover of V is a combination J of routes $j \in R$, where each customer $i \in V$ is covered by at least one route $j \in R$. Let c_j be the cost of each route j and the binary constant ρ_{ij} that informs if customer i is served by route j. The Set Covering problem (SCP) consists in finding a set $J \subseteq R$ such that the total cost is minimized.

In order to present the mathematical model for the SCP, let y_j be a binary variable associated with a route $j \in J$. Follow the model used in our algorithm.

$$\min \sum_{j \in R} c_j y_j$$

Subject to:

$$\sum_{j \in R} \rho_{ij} y_j \geq 1, \qquad \forall i \in V \qquad (1)$$

$$\sum_{j \in R} y_j \leq |K|, \qquad (2)$$

$$y_j \in \{0, 1\} \qquad (3)$$

5   Computational Experiments

The developed heuristic was coded in C/C++ using the mathematical solver Gurobi 8.0.0, with all its default settings. All experiments were performed on an Intel Core i7-7700K processor with 3.60 GHz and 16 GB RAM running Ubuntu 16.04 LTS 64 bits.

5.1   Instances Description

In order to validate our approaches, the computational experiments were performed using three groups of instances (E, I, U) proposed by Parragh [19]. These instances have heterogeneity based on data given by the Austrian Red Cross. The instances were randomly generated in the square $[-10, 10]^2$ using the uniform distribution. The depot was located in the center of the plan. Each edge $(v_i, v_j) \in A$ has the cost c_{ij} and travel time t_{ij} were calculated using Euclidean distance. Each vertex i was defined a time window $[e_i, l_i]$. The pickup vertices e_i were established in a range of $[0, T - 60]$, where T is the time of planning horizon and l_i was set as $e_i + 15$. The delivery vertices l_i were defined in the interval $[60, T]$ and e_i was set as $l_i - 15$. Each instance is represented by the name ak-n, where k is the number of vehicles used and n is the total number of requests.

In group E of instances, half of the patients are normal ones, 25% are wheelchair ones and 25% stretcher ones. 10% of the patients have one companion. The fleet of vehicles for this group is homogeneous with vehicles of type V1. Each vehicle V1 has two places to companions, one for a wheelchair patient, one for the stretcher patient and one for a normal patient. In group I of instances, 83% of requests are from normal patients, 11% are wheelchair ones, and 6% are stretcher ones. Half of the patients have one companion. The fleet of vehicles is heterogeneous with vehicles of types V1 and V2. Each vehicle V2 has one place to companions, one for a wheelchair patient, one for stretcher patient and six for the normal patients. In group U of instances, are just considered the normal patients and there are no companions. The fleet of vehicles is homogeneous with vehicles of type V3. Each vehicle V3 just has places for normal patients. Table 2 shows the heterogeneity of patients and vehicles by instance groups (Table 1).

Table 1. Patients occupation

Patient type	Place type:			
	Normal	Wheelchair	Stretcher	Companion
Normal	x	x		
Wheelchair		x		
Stretcher			x	
Companion	x	x		x

5.2   Experimental Results

For each instance, the algorithm was executed 10 times using as stop criteria $iterMax = 100$, which represents the maximum number of the iterations without improving the

Table 2. Instances information

Instance group	Probability of patient be:				Fleet of vehicles
	Normal	Wheelchair	Stretcher	Companion	
E	0.50	0.25	0.25	0.10	hom(V1)
I	0.83	0.11	0.06	0.50	het(V1, V2)
U	1.00	0.00	0.00	0.00	hom(V3)

hom = Homogeneous, het = Heterogeneous
V1: 2 Companions, 1 Normal, 1 Wheelchair, 1 Stretcher
V2: 1 Companion, 6 Normal, 1 Wheelchair, 1 Stretcher
V3: 0 Companion, 3 Normal, 0 Wheelchair, 0 Stretcher

best solution found. For running the set covering problem was used half of the $iterMax$ iterations. The parameters used in the algorithm were calibrated using the IRACE package [13]. Table 3 shows the tested configurations for the parameters. The best configurations returned by IRACE are boldfaced. The neighborhoods are represented as follow and the selected neighborhoods are presented in Sect. 4.3:

1 - Relocation - Relocation of one request of one vehicle to another.
2 - Swap - Exchange of one request of one vehicle to another.
3 - Crossover - Crossover between two vehicles.
4 - Swap(2) - Exchange of two requests of one vehicle to another.
5 - Swap(2-1) - Exchange of two requests of one vehicle with one request from another.
6 - Relocation(2) - Relocation of two requests of one vehicle to another.

Table 3. IRACE Calibration

Parameter	Values
Local search	123, 124, 125, 126, 134, 135, 136, 145, 146, 156
Neighborhood	**234**, 235, 236, 245, 246, 256, 345, 346, 356, 456
Initial penalty	1, 3, 5, 8, **10**
iterMax	50, **100**, 150

Table 4 presents the results on the instances group U, E and I. In group U, the customers and vehicle fleet were considered as homogeneous. In group E the customers were considered heterogeneous and the vehicle fleet was homogeneous. In the group I the customers and vehicle fleet are heterogeneous. In this table, the column **Instance** represents the name for each instance, this name is composed by the number of vehicles used and the number of requests. The columns **Parragh** [19] and **Braekers et al.** [3] show the results obtained by these authors in their experiments. The column **Best VNS** presents the best solution found by our algorithm and the column **AVG VNS** shows the average values in 10 executions of the VNS algorithm. For each column of the table, **sol** reports the objective function value, **gap** the difference between the solution value found and the best-known solution (BKS) value, while the column **time** shows the time spend by the algorithm in seconds. The times in this work are from different

Table 4. Results for instances U, E and I

Instances	Parragh [19]			Braekers et al. [3]			Best VNS		AVG VNS		
	Sol.	Gap%	Time	Sol.	Gap%	Time	Sol.	Gap%	Sol.	Gap%	Time
Group U											
a2−16	**294.25**	0.00	68.20	**294.25**	0.00	8.60	**294.25**	0.00	**294.25**	0.00	3.83
a2−20	**344.83**	0.00	133.80	**344.83**	0.00	20.20	**344.83**	0.00	**344.83**	0.00	15.69
a2−24	**431.12**	0.00	187.80	**431.12**	0.00	17.40	434.53	0.78	436.48	1.23	42.26
a3−18	**300.48**	0.00	45.40	**300.48**	0.00	9.40	**300.48**	0.00	302.53	0.68	2.60
a3−24	**344.83**	0.00	86.80	**344.83**	0.00	16.60	344.91	0.02	347.50	0.77	14.58
a3−30	**494.85**	0.00	105.60	**494.85**	0.00	18.80	500.51	1.13	502.52	1.53	74.02
a3−36	583.30	0.02	162.60	**583.19**	0.00	28.40	599.43	2.71	607.24	3.96	247.54
a4−16	**282.68**	0.00	26.00	**282.68**	0.00	9.80	283.10	0.15	283.10	0.15	1.04
a4−24	**375.02**	0.00	50.80	**375.02**	0.00	13.00	379.36	1.14	380.81	1.52	6.36
a4−32	486.88	0.28	86.00	**485.50**	0.00	25.60	487.31	0.37	491.70	1.26	54.24
a4−40	561.80	0.74	130.60	**557.69**	0.00	26.40	**557.69**	0.00	569.36	2.05	243.66
a4−48	673.64	0.72	253.80	**668.82**	0.00	35.40	678.98	1.50	683.13	2.09	580.92
Group E											
a2−16	**331.16**	0.00	65.60	**331.16**	0.00	9.40	**331.16**	0.00	**331.16**	0.00	4.89
a2−20	**347.03**	0.00	120.00	**347.03**	0.00	19.60	**347.03**	0.00	**347.03**	0.00	16.32
a2−24	**450.25**	0.00	160.40	**450.25**	0.00	15.80	450.38	0.02	452.92	0.65	76.08
a3−18	**300.63**	0.00	47.60	**300.63**	0.00	9.60	**300.63**	0.00	302.01	0.46	2.05
a3−24	**344.91**	0.00	76.20	**344.91**	0.00	14.60	**344.91**	0.00	347.63	0.86	17.73
a3−30	**500.58**	0.00	107.60	**500.58**	0.00	17.00	505.64	1.00	507.72	1.54	108.75
a3−36	**583.19**	0.00	161.60	**583.19**	0.00	23.60	599.43	2.71	610.46	4.47	316.76
a4−16	**285.99**	0.00	25.00	**285.99**	0.00	8.20	291.55	1.91	291.76	1.98	1.25
a4−24	**383.84**	0.00	52.60	**383.84**	0.00	12.20	386.06	0.58	289.09	1.35	8.60
a4−32	502.52	0.45	83.00	**500.24**	0.00	22.80	501.85	0.32	507.11	1.35	49.16
a4−40	585.64	0.90	121.00	**580.42**	0.00	24.20	589.29	1.51	597.89	2.92	204.60
a4−48	675.37	0.72	252.20	**670.52**	0.00	33.60	676.28	0.85	688.57	2.62	786.57
Group I											
a2−16	**294.25**	0.00	68.40	**294.25**	0.00	7.20	**294.25**	0.00	**294.25**	0.00	5.33
a2−20	**355.74**	0.00	141.80	**355.74**	0.00	17.40	360.23	1.25	362.69	1.92	14.16
a2−24	**431.12**	0.00	211.00	**431.12**	0.00	12.60	434.53	0.78	441.32	2.31	59.08
a3−18	**302.17**	0.00	47.20	**302.17**	0.00	8.40	**302.17**	0.00	303.27	0.36	3.53
a3−24	**344.83**	0.00	83.60	**344.83**	0.00	13.40	345.31	0.14	350.79	1.70	19.23
a3−30	**494.85**	0.00	106.80	**494.85**	0.00	14.80	502.77	1.57	511.75	3.30	93.29
a3−36	618.58	0.07	170.60	**618.15**	0.00	22.60	636.97	2.95	641.98	3.71	296.38
a4−16	**299.05**	0.00	27.00	**299.05**	0.00	7.20	302.87	1.26	307.23	2.66	1.09
a4−24	375.07	0.01	51.60	**375.02**	0.00	12.00	379.36	1.14	381.91	1.80	9.83
a4−32	**486.93**	0.00	88.00	**486.93**	0.00	21.00	488.74	0.37	496.25	1.88	59.96
a4−40	561.35	0.66	132.20	**557.69**	0.00	23.80	566.11	1.49	573.02	2.68	279.11
a4−48	680.43	1.45	262.40	**670.72**	0.00	30.00	684.37	1.99	695.95	3.63	749.48

computers, the computer used by Parragh [19] was an Intel Pentium D with 3.20 GHz and 4 GB RAM and the computer used by Braekers et al. [3] was an Intel Core with 2.6 GHz and 4 GB RAM, and the computer used in this work is an Intel Core I7-7700 CPU @ 3.60 GHz with 16 GB RAM. Some methods have been developed for a fair comparison, methods that measure the performance of computers in general, and can be found in work of Dongarra [7].

According to Table 4, we can see that our algorithm was able to find the BKS solution for 10 of 36 instances, where they are four in group U and E and two in group I. The VNS found solutions with gap up to 1% for 12 of 36 cases, where they are four in group U, five in group E and three in group I. In the last 14 cases our algorithm obtained solutions with gap up to 2.95%.

Regarding computation time, our algorithm on average is faster in 26 out of 32 instances compared to Parragh [19] and find the better solution just for one case (a4−40 of group U). Compared to Braekers et al. [3], the VNS algorithm on average is faster just for half of 32 instances.

6   Conclusions

In this paper, we propose a simple heuristic algorithm based on the Variable Neighborhood Search for the Dial-a-Ride Problem. This problem is a variation of the Vehicle Routing Problem, where the customer's convenience is taken into account. We considered a heterogeneous demand for customers and vehicles. Our algorithm was tested in the instances described in the literature. In three groups with a total of 36 instances. The VNS algorithm showed able to find 10 of 36 best-known solutions, and for 12 of 36 solutions, we find a solution with a gap less than 1%. For the other 14 instances, in the worst case, we find a solution with a gap of 2.95%.

Acknowledgments. The authors thank Coordenação de Aperfeiçoamento de Pessoal de Nível Superior (CAPES), Fundação de Amparo à Pesquisa do Estado de Minas Gerais (FAPEMIG), Conselho Nacional de Desenvolvimento Científico e Tecnológico (CNPq), Universidade Federal de Ouro Preto (UFOP) and Universidade Federal de Viçosa (UFV) for supporting this research.

References

1. Attanasio, A., Cordeau, J.F., Ghiani, G., Laporte, G.: Parallel Tabu search heuristics for the dynamic multi-vehicle dial-a-ride problem. Parallel Comput. **30**, 231–236 (2004)
2. Beaudry, A., Laporte, G., Melo, T., Nickel, S.: Dynamic transportation of patients in hospitals. OR Spectr. **32**, 77–107 (2010). https://doi.org/10.1007/s00291-008-0135-6
3. Braekers, K., Caris, A., Janssens, G.K.: Exact and meta-heuristic approach for a general heterogeneous dial-a-ride problem with multiple depots. Transp. Res. Part B Methodol. **67**, 166–186 (2014)
4. Cordeau, J.F., Laporte, G.: A Tabu search heuristic for the static multi-vehicle dial-a-ride problem. Transp. Res. Part B Methodol. **37**, 579–594 (2003)
5. Cordeau, J.F.: A branch-and-cut algorithm for the dial-a-ride problem. Oper. Res. **54**, 573–586 (2006)

6. Cordeau, J.F., Laporte, G.: The dial-a-ride problem: models and algorithms. Ann. Oper. Res. **153**, 29–46 (2007). https://doi.org/10.1007/s10479-007-0170-8

7. Dongarra, J.J.: Performance of various computers using standard linear equations software. ACM SIGARCH Comput. Arch. News **20**, 22–44 (2014)

8. Hanne, T., Melo, T., Nickel, S., Melo, T., Nickel, S.: Bringing robustness to patient flow management through optimized patient transportation in hospitals. Interfaces **39**, 241–255 (2018)

9. Hansen, P., Mladenović, N.: Variable neighborhood search: principles and applications. Eur. J. Oper. Res. **130**, 449467 (2001)

10. Ho, S.C., Szeto, W.Y., Kuo, Y.H., Leung, J.M.Y., Petering, M., Tou, T.W.H.: A survey of dial-a-ride problems: literature review and recent developments. Transp. Res. Part B Methodol. **111**, 395–421 (2018)

11. Jaw, J.-J., Odoni, A.R., Psaraftis, H.N., Wilson, N.H.M.: A heuristic algorithm for the multi- vehicle advance request dial-a-ride problem with time windows. Transp. Res. **20B**, 243–257 (1986)

12. Jorgensen, R.M., Larsen, J., Bergvinsdottir, K.B.: Solving the dial-a-ride problem using genetic algorithms. J. Oper. Res. Soc. **58**, 1321–1331 (2007)

13. López-Ibáñez, M., Dubois-Lacoste, J., Cáceres, L.P., Birattari, M., Stützle, T.: The irace package: Iterated racing for automatic algorithm configuration. Oper. Res. Perspect. **3**, 43–58 (2016)

14. Lutz, W., Sanderson, W., Scherbov, S.: The coming acceleration of global population ageing. Nature **451**, 716–719 (2008)

15. Madsen, O.B.G., Ravn, H.F., Rygaard, J.M.: A heuristic algorithm for a dial-a-ride problem with time windows, multiple capacities, and multiple objectives. Ann. Oper. Res. **60**, 193–208 (1995). https://doi.org/10.1007/BF02031946

16. Masmoudi, M.A., Braekers, K., Masmoudi, M., Dammak, A.: A hybrid genetic algorithm for the heterogeneous dial-a-ride problem. Comput. Oper. Res. **81**, 1–13 (2017)

17. Mauri, G.R., Antonio, L., Lorena, N.: Customers' satisfaction in a dial-a-ride problem. IEEE Intell. Transp. Syst. Mag. **3**, 6–14 (2009)

18. Parragh, S.N., Doerner, K.F., Hartl, R.F.: Variable neighborhood search for the dial-a-ride problem. Comput. Oper. Res. **37**, 1129–1138 (2010)

19. Parragh, S.N.: Introducing heterogeneous users and vehicles into models and algorithms for the dial-a-ride problem. Transp. Res. Part C Emerg. Technol. **19**, 912–930 (2011)

20. Parragh, S.N., Schmid, V.: Hybrid column generation and large neighborhood search for the dial-a-ride problem. Comput. Oper. Res. **40**, 490–497 (2013)

21. Schilde, M., Doerner, K.F., Hartl, R.F.: Metaheuristics for the dynamic stochastic dial-a-ride problem with expected return transports. Comput. Oper. Res. **12**, 1719–1730 (2011)

22. Toth, P., Vigo, D.: The Vehicle Routing Problem. SIAM Monographs on Discrete Mathematic and Applications, Philadelphia (2002)

Multi-objective Basic Variable Neighborhood Search for Portfolio Selection

Thiago Alves de Queiroz[1]([⊠]) [iD], Leandro Resende Mundim[2],
and André Carlos Ponce de Leon Ferreira de Carvalho[2] [iD]

[1] Institute of Mathematics and Technology, Federal University of Goiás,
Campus Catalão, Catalão, GO 75704-020, Brazil
taq@ufg.br
[2] Institute of Mathematics and Computer Sciences, University of São Paulo,
São Carlos, SP 13566-590, Brazil
{mundim,andre}@icmc.usp.br

Abstract. The Portfolio Selection Problem looks for a set of assets with
the best trade-off between return and risk, that is, with the maximum
expected return and the minimum risk (e.g., the variance of returns).
As these objectives are conflicting, it is a difficult multi-objective problem. Different models and algorithms have been proposed to obtain the
(optimal) Pareto front. However, exact approaches take days for a large
set of points to the Pareto front. Within this perspective, we develop
a basic variable neighborhood search heuristic to solve the bi-objective
portfolio selection problem. The proposed heuristic considers ten neighborhood structures that are mainly based on swap moves and has a local
improvement based on averaging the proportions that are invested in
consecutive assets. The proposed heuristic was experimentally compared
with the Mean-Variance model of Markowitz, using benchmark instances
from the OR-Library. The number of assets in these instances ranges from
31 to 225. According to the experimental results, the proposed heuristic
performed well in the construction of different Pareto fronts.

Keywords: Portfolio optimization · Basic variable neighborhood
search · Mean-variance model · Multiobjective optimization

1 Introduction

According to the modern portfolio theory [17,18], it is possible to select portfolios with the minimum risk for a given expected return. Portfolios can also be
selected by maximizing the expected return for a given level of risk [6]. Aiming
at analyzing the trade-off between risk and expected return, the Mean-Variance
(MV) model in [17], which is bi-objective and quadratic, can be used to obtain
a front of optimal portfolios in terms of minimum risk, which is measured as the
variance of returns, and maximum expected return.

© Springer Nature Switzerland AG 2020
R. Benmansour et al. (Eds.): ICVNS 2019, LNCS 12010, pp. 67–80, 2020.
https://doi.org/10.1007/978-3-030-44932-2_5

Recently, the survey in [11] discussed modelings and applications of the MV model and its variants, showing the importance of this model in solving the portfolio selection problem. The authors reviewed 175 papers that were published in the last two decades, precisely from 1998 to 2018. From this survey, we notice that the number of published articles is significantly higher in the last decade (i.e., from 2009 to 2018), with 136 published papers out of 175. Despite several attempts to linearize the measure of risk in the MV model, as the Mean Absolute Deviation (MAD) model in [13] and the Gini Mean difference (GMD) model in [25], the literature has been solving it with heuristic methods.

In terms of heuristics for the MV model, in [9] there is a multi-objective evolutionary algorithm (MOEA), where cardinality constraints (i.e., there is a limit on the number of assets that can be invested) are included. The results of the MOEA were compared with those of the resolution of the MV model, showing that for small values of cardinality (i.e., about five assets), both attained similar results. In [21], there is another MOEA in which three-objectives are considered, that is, to maximize the expected return, to minimize the uncertainty risk, and to minimize the relation risk. The binary codification was adopted to the chromosomes and indicates whether an asset is in the portfolio. The crowding distance was used to sort the chromosomes and to determine the Pareto front. The computational results showed that such a method could give flexible and accurate Pareto fronts.

Different heuristics were implemented in [24] for the MV model with cardinality, floor, and round-lot constraints. Some of the heuristics are the Vector Evaluated Genetic Algorithm (VEGA) from [23], the Multi-objective Optimization Genetic Algorithm (MOGA) from [12], the Strength Pareto Evolutionary Algorithm second version (SPEA2) from [26], and the Non-dominated Sorting Genetic Algorithm II (NSGA-II) from [7]. The computational results showed the NSGA-II and SPEA2 could produce diverse portfolios in the Pareto front, as well as they had similar results, although the SPEA2 performed a little better.

In [1], the Pareto Envelope-based Selection Algorithm (PESA) from [5], SPEA2, and NSGA-II are codified for a variant of MV model with three objectives, to maximize the expected return, to minimize the variance of returns, and to minimize the number of assets in the portfolio. The computational results indicated that the SPEA2 had the best performance in terms of Pareto fronts, whereas the PESA was the fastest one. In [2], the authors compared five multi-objective evolutionary-based heuristics for the MV model with cardinality constraints. In such a study, the NSGA-II and SPEA2 outperformed the others in terms of solution quality and convergence criteria to the optimal Pareto front. The authors noticed that the SPEA2 could perform even better than NSGA-II for some instances.

The NSGA-II and SPEA2 were also used in [15] and [16]. In the recent work of [16], instead of the MV model, the authors solved a Mean-Semivariance model, which considers only adverse return variations, by using the NSGA-II and SPEA2. In such work, the NSGA-II performed better in comparison with the SPEA2, although both achieved Pareto fronts with good outcomes in terms of Bollinger bands. Although we could not find any work in which there was a clear superiority of the NSGA-II over the SPEA2, the NSGA-II seems to be more popular, as pointed in [11]. These authors listed 26 papers that used the NSGA and NSGA-II and 14 ones that used the SPEA2 for solving the MV model and its variants.

Recent works have also considered other heuristic frameworks, not based on evolutionary algorithms, to successfully solve portfolio selection problems, as swarm optimization and the variable neighborhood search. In [14], a covariance guided artificial bee colony algorithm is proposed to the MV model. Then, NSGA-II is used when ranking and generating non-dominated solutions. In [4], there is a multi-objective particle swarm optimization to solve a variant of the MV model in which higher moments (i.e., based on skewness and kurtosis) are included. On the other hand, in [22], there is a variable neighborhood search based heuristic for solving large instances of the project portfolio selection problem with uncertainty. The authors aimed at maximizing the expected return, while considering random cash flows, discount rates, and a given level of risk.

With this in mind, we propose a multi-objective basic variable neighborhood search [8], to solve the MV model and build an effective Pareto front of portfolios. The proposed heuristic considers ten neighborhood structures mainly based on swap movements. Each portfolio is represented as a vector of assets, where each position of the vector has the proportion of the total capital that is invested in the respective asset. The remainder of this work is organized as follows: Sect. 2 describes the problem and the MV model; Sect. 3 has the proposed heuristic for the problem; Sect. 4 presents the computational experiments on instances from the OR-Library, where a comparison is made between the solutions of the proposed heuristic and those from the MV model; and, Sect. 5 has the concluding remarks and proposals for future works.

2   Problem Formulation

In the Portfolio Selection Problem (PSP), there are n assets, each one with an expected return per period μ. The covariance between two assets is represented by σ. Let $x_i \geq 0$ indicates the proportion of the total capital that is invested in the asset i. Then, the objective of the problem is to determine the percentage to invest in each asset i, for $i = 1, \ldots, n$, to achieve a portfolio with the maximum expected return and the minimum risk.

In [17], the PSP is tackled employing a mean-variance model in which the risk is measured according to the variance of the portfolio's expected return. With the non-negative variables x_i, such model is then defined in (1)–(4).

In the MV model (1)–(4), there are two objectives, that is, to maximize the expected return (1) and to minimize the variance of the expected return (2), which is quadratic in the variables x. These objectives are conflicting since it is usually well-documented that the smaller the variance, the lower the expected return. Constraint (3) ensures the proportions that are invested in all assets must sum up to one, while the domain of variables is expressed in (4).

$$\text{Maximize} \quad E = \sum_{i=1}^{n} \mu_i x_i \tag{1}$$

$$\text{Minimize} \quad V^2 = \sum_{i=1}^{n} \sum_{j=1}^{n} \sigma_{ij} x_i x_j \tag{2}$$

$$\text{Subject to:} \quad \sum_{i=1}^{n} x_i = 1, \tag{3}$$

$$x_i \geq 0, \qquad\qquad \forall\, i = 1, \ldots, n. \tag{4}$$

3   A Heuristic for the Portfolio Selection Problem

The Variable Neighborhood Search (VNS) [19] is a single solution based heuristic that has been showing powerful to tackle different mono- and multi-objective optimization problems [8,10]. In the VNS, the solution travels through neighborhood structures to become globally optimal in relation to all neighborhoods. When an improved solution is found, the VNS comes back to the first neighborhood, or else it continues toward the next neighborhood.

In [8], the VNS is adapted to solve multi-objective optimization problems. These authors discussed the multi-objective versions of the reduced VNS, the variable neighborhood descent, and the general VNS, pointing out the advantages of each one. The reduced VNS has the particularity of obtaining new solutions by random perturbations, while the variable neighborhood descent considers local searches with deterministic perturbations. The general VNS combines random perturbations with local searches that are related to the variable neighborhood descent to balance diversification and intensification of the search.

Another version that is not discussed in [8] is the basic VNS, where local searches do not need to consider the variable neighborhood descent. Based on this, we develop a Multi-Objective Basic VNS (MO-BVNS) for the PSP for which the following ten neighborhood structures are proposed.

In the MO-BVNS, a solution represents an approximated Pareto front of the PSP. The Pareto front is a set of efficient (i.e., non-dominated) points p_x. In each p_x, x is a vector of size n and x_i holds the proportion of the total capital that is invested in the asset i, for $i = 1, \ldots, n$. Figure 1 illustrates examples of vectors x from points p_x considering six assets. A point p_x dominates a point p'_x if the expected return (see (1)) and the risk (see (2)) of p_x is, respectively, greater and smaller than those of p'_x. Therefore, a point is non-dominated by any other of the Pareto front if one of its objective (expected return or risk) value is better than the respective objective of the other points. The proposed neighborhood structures are discussed next, where Fig. 1 has an example of a neighbor point of p_x considering each of the neighborhood structures:

- Neighborhood N1: select two assets i and j, and swap the values in x_i and x_j;
- Neighborhood N2: select two assets i and j, and define x_i and x_j as $\frac{x_i + x_j}{2}$ (i.e., they receive the average of their values);
- Neighborhood N3: select two assets i and j, and define x_i as $x_i + x_j$ and x_j as 0 (i.e., the proportion of j is added into i);
- Neighborhood N4: divide the assets in two sets A and B, where A contains the $\frac{n}{2}$ assets with the highest expected returns and B the remaining assets. Select two assets, i in A and j in B, and define x_i as $x_i + x_j$ and x_j as 0;
- Neighborhood N5: divide the assets in two sets C and D, where C contains the $\frac{n}{2}$ assets with the lowest risks and B the remaining assets. Select two assets, i in C and j in D, and define x_i as $x_i + x_j$ and x_j as 0;
- Neighborhood N6: consider the two sets C and D as in N5, and select two assets, i in C and j in D. Define x_i and x_j as $\frac{x_i + x_j}{2}$;
- Neighborhood N7: let E be the set with all assets i for which $x_i \leq \beta$ (i.e., E contains all assets whose the proportion invested in each one is less than β percent of the total capital). Define s as $\frac{\sum_{i \in E} x_i}{\sum_{j \notin E} 1}$, x_i as 0 for each $i \in E$, and update the remaining assets x_j as $x_j + s$, for each $j \notin E$;
- Neighborhood N8: let i be the asset in which $x_i \geq \gamma$ (i.e., the proportion invested in i is more than γ percent of the total capital). Define s as $\frac{x_i}{n-1}$, x_i as 0, and update the remaining assets x_j as $x_j + s$, for each $j = 1, \ldots, n$ and $j \neq i$;
- Neighborhood N9: select two assets i and j, such that $i < j$, and define the set F as $\{i, i+1, \ldots, j\}$. Define s as $\frac{\sum_{k \in F} x_k}{\sum_{l \notin F} 1}$, x_k as 0 for each $k \in F$, and update the remaining assets x_l as $x_l + s$, for each $l \notin F$;
- Neighborhood N10: select two assets i and j, such that $i < j$, and define the set F as $\{i, i+1, \ldots, j\}$. Define s as $\frac{\sum_{k \notin F} x_k}{|F|}$, x_k as 0 for each $k \notin F$, and update the remaining assets x_l as $x_l + s$, for each $l \in F$.

Algorithm 1 describes the proposed MO-BVNS that has the neighborhood structures N1 to N10. As input, it receives the total number of neighborhood structures k_{max}, which is ten, the number of consecutive assets na that are combined in the local improvement phase, and the time limit that is used as the stopping criterion. As output, it returns the approximated Pareto front E_{best}.

The heuristic starts with a solution E whose points p_x are generated from randomly dividing the total capital among the n assets, for a total of 500 points. After creating these points, only the non-dominated ones are kept. While the imposed time limit is not reached, the heuristic considers an inner-loop (lines 5–9) that iterates over the neighborhood structures by considering a perturbation phase (i.e., MO-Shake), a local improvement phase (i.e, MO-LocalImp), and a neighborhood change phase (i.e., MO-NeighborhoodChange).

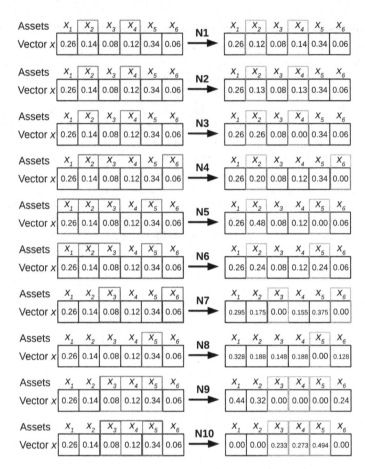

Fig. 1. Example of neighbor points considering each of the neighborhood structures.

Algorithm 1: MULTI-OBJECTIVE BASIC VARIABLE NEIGHBORHOOD
SEARCH FOR THE PORTFOLIO SELECTION PROBLEM

    Input: K, na, T.
    Output: E_{best}.
1 $E \leftarrow$ randomly generated solution;
2 $E_{best} \leftarrow E$;
3 **do**
4 | $k \leftarrow 1$;
5 | **do**
6 | | $E' \leftarrow$ MO-Shake(E, k);
7 | | $E'' \leftarrow$ MO-LocalImp(E', na);
8 | | MO-NeighborhoodChange(E, E'', E_{best}, k);
9 | **while** $k \leq k_{max}$;
10 **while** $t \leq t_{max}$;
11 **return** E_{best};

The MO-Shake phase is presented in Algorithm 2. In this phase, a new solution is obtained from E by considering the k-th neighborhood structure. For each point p_x in E, a new point p'_x is randomly (given a uniform distribution) obtained by applying the neighborhood Nk. The new points originate a new solution, E', which is the output of this phase.

Algorithm 2: MO-SHAKE.

Input: E, k.
Output: E'.
1 $E' \leftarrow \emptyset$;
2 **foreach** $p_x \in E$ **do**
3     $p'_x \leftarrow$ apply the neighborhood Nk to p_x;
4     $E' \leftarrow E' \cup \{p'_x\}$;
5 **end**
6 **return** E';

The MO-LocalImp phase is described in Algorithm 3. It consists of generating new points from averaging two consecutive points of a given solution E. It has been observed that non-dominated points can emerge from combining the proportions that are invested in two consecutive assets. The number of successive assets that are combined is due to the parameter $na \leq n$.

Algorithm 3: MO-LOCALIMP.

Input: E, na.
Output: E'.
1 $E' \leftarrow E$;
2 **foreach** $p_x \in E$ **do**
3   **for** $i \leftarrow 1, \ldots, na - 1$ **do**
4        $m \leftarrow \frac{x_i + x_{i+1}}{2}$;
5        $x_i \leftarrow m$;
6        $x_{i+1} \leftarrow m$;
7   **end**
8     $E' \leftarrow E' \cup \{p_x\}$;
9 **end**
10 **return** E';

In the MO-NeighborhoodChange phase, see Algorithm 4, the current solution E and the best solution E_{best} are updated if the solution E'', which is obtained after the local improvement phase, has at least one non-dominated point in comparison with E. If this is the case, then the current and best solutions are updated to contain the new non-dominated points, and the search is reset to the first neighborhood. Otherwise, the search continues with E to the next neighborhood structure.

Algorithm 4: MO-NEIGHBORHOODCHANGE.

Input: E, E', E_{best}, k.

1 **if** *MO-Improvement(E, E')* **then**
2     $E_{best} \leftarrow E_{best} \cup E' \cup E$;
3   Remove all dominated points of E_{best};
4     $E \leftarrow E_{best}$;
5     $k \leftarrow 1$;
6 **else**
7   Remove all dominated points of E;
8     $k \leftarrow k + 1$;
9 **end**

Algorithm 5 describes the routine MO-Improvement. This routine aims at checking whether the current solution E' contains at least one non-dominated point in comparison with all the points in E. If this is true, then the current solution E is updated to contain this new non-dominated point (and possibly others), according to Algorithm 4.

Algorithm 5: MO-IMPROVEMENT.

Input: E, E'.
Output: *True* or *False*.

1 **foreach** $p_x \in E'$ **do**
2   **if** $p_x \notin E$ **AND** *non-dominated*(p_x, E) **then**
3   **return** *True*;
4   **end**
5 **end**
6 **return** *False*;

---

4 Computational Experiments

All the algorithms were coded in the C++ programming language and the experiments were carried out in a computer with Intel® Core^{TM} i7-2600, 3.40 GHz, 16 GB of RAM, and Ubuntu 12.04 LTS. A time limit of 60 s was imposed on solving each instance as the stop criterion, and the proposed heuristic was executed five times (with different seeds). Regarding the parameters of the heuristic, they were calibrated by trial and error, choosing after all na as the total number of assets n, β as 0.1, and γ as 0.9.

The authors in [11] discussed the most used data sets, pointing out that about 2% of the literature has adopted hypothetical data sets, 39% of it have used instances from the OR-Library, and 59% of it have used data sets of case studies. Hypothetical data sets have the inconvenient of hardly representing the reality of stock markets, while data sets of case studies are not frequently available for other researchers and practitioners. Therefore, the instances in this work were obtained from the OR-Library [3], and Table 1 has details of them.

Table 1. Data sets that are used in this work.

Name	Index	Country	Number of assets
Port1	Hang Seng	Hong Kong	31
Port2	DAX	Germany	85
Port3	FTSE	United Kingdom	89
Port4	S&P	United States	98
Port5	Nikkei	Japan	225

To assess the quality of the Pareto fronts that are obtained by the proposed heuristic, we solve the MV model in (1)–(4) by the approach in Algorithm 6. This approach is based on the ϵ-constraint method, and it was used in [3] to compare the results of heuristics with a near-optimal Pareto front. The idea is to solve a mono-objective MV model, that is, to maximize the expected return (1), where the other objective, to minimize the risk (variance of returns) (2), is transformed into a constraint that is limited by a given ρ. Then, the Pareto front is composed of non-dominated points that are solutions of the mono-objective MV model for different values of ρ. Each mono-objective model is solved with a sequential least squares programming algorithm that uses the Han-Powell quasi-Newton method [20].

Algorithm 6: APPROACH FOR A MONO-OBJECTIVE MV MODEL.

Input: n, μ, σ, N, $[\rho_{min}, \rho_{max}]$.
Output: Pareto front F.

1 $F \leftarrow \emptyset$;

2 $S \leftarrow \frac{\rho_{max} - \rho_{min}}{N}$;

3 **for** $i = 1, \ldots, N$ **do**

4 $\quad \rho \leftarrow \rho_{min} + i \times S$;

5 $\quad x \leftarrow$ with the ϵ-constraint method, to solve the MV model (1), (3),
$\quad$ (4), and $\sqrt{\sum_{i=1}^{n} \sum_{j=1}^{n} \sigma_{ij} x_i x_j} \leq \rho$;

6 $\quad F \leftarrow F \cup \{x\}$;

7 **end**

8 Remove all dominated points of F;

9 **return** F;

In Algorithm (6), the input has information of the instance, as the number of assets n, the expected return μ_i of each asset i, the covariance σ_{ij} between assets i and j, and the quantity N of values of ρ that is considered. The values of ρ are obtained from the interval $[\rho_{min}, \rho_{max}]$, where ρ_{min} is function (2) calculated with the solution of the linear MV model with the objective function (1) set as *minimize* and constraints (3)–(4). On the other hand, ρ_{max} is function (2) calculated with the solution of the linear MV model with the objective function (1) set as *maximize* and constraints (3)–(4). Notice that in these linear models, the second objective (i.e., related to the risk) is not considered, and its function is used to determine the interval in which the risk varies in the ϵ-constraint method. In the experimental tests, we imposed a time limit of 48 h, and then, for instances Port1, Port2, and Port3, we obtained $N = 1000$, while for instances Port4 and Port5, we obtained $N = 100$.

The Pareto fronts of the MO-BVNS and the approach for the MV model in Algorithm 6 are illustrated in Figs. 2, 3, 4, 5 and 6, for the respective instances Port1 to Port5. The dashed curve in red (i.e., Pareto F) represents the Pareto front of the approach in Algorithm 6, while the others are the Pareto fronts obtained with the MO-BVNS from each of the five runs (different seeds).

Observing Figs. 2, 3, 4, 5 and 6, the MO-BVNS has obtained Pareto fronts that differ from each other, especially in the instances that have more assets, Port4 and Port5. However, the differences between them are quite small, showing a low variability in the results of the heuristic when different seeds are used. Concerning the solution of MV model, the Pareto F has fewer points in comparison with the Pareto fronts of the MO-BVNS. We notice the left part (along the direction of the risk, on x-axis) of the Pareto F dominates the left part of the Pareto fronts of the heuristic, for all the instances. On the other hand, the right part of the Pareto fronts of the heuristic has many non-dominated points (i.e., they dominate) in comparison with the Pareto F, especially in the instances Port2, Port3, Port4, and Port5.

In the results of Figs. 2, 3, 4, 5 and 6, we observe the MO-BVNS has generated Pareto fronts with a large number of points that are better spread and spaced. On the other hand, notice that the Pareto F is quite small (i.e., it has few points)

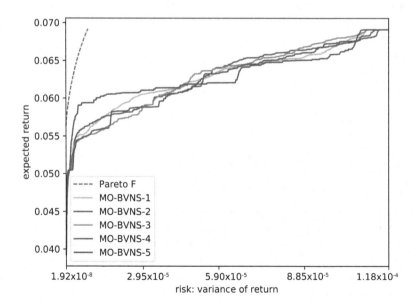

Fig. 2. Pareto fronts for Port1. (Color figure online)

for the instances Port2, Port3, and Port4. It is worth mentioning the Pareto fronts of the heuristic have confirmed the higher the risk, the higher the expected return.

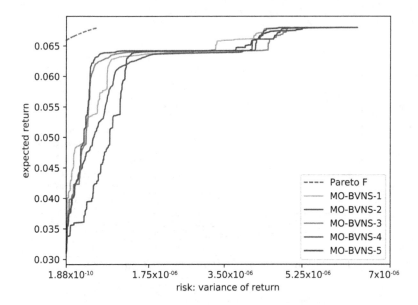

Fig. 3. Pareto fronts for Port2. (Color figure online)

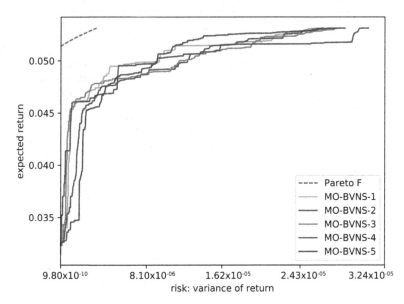

Fig. 4. Pareto fronts for Port3. (Color figure online)

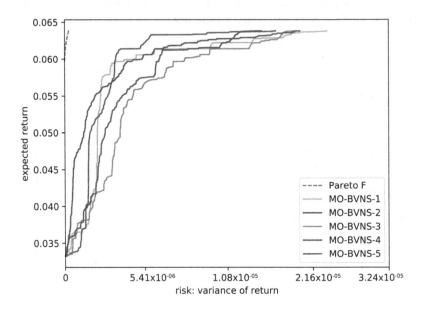

Fig. 5. Pareto fronts for Port4. (Color figure online)

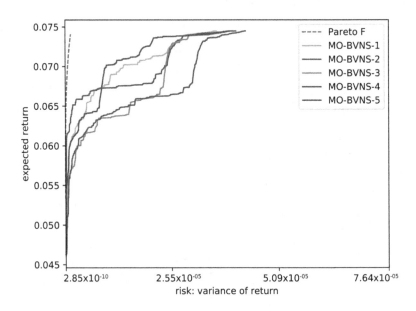

Fig. 6. Pareto fronts for Port5. (Color figure online)

5   Concluding Remarks

In this work, a bi-objective portfolio selection problem is solved with a basic variable neighborhood search based heuristic, namely the MO-BNVS. The proposed heuristic considers ten neighborhood structures, with movements that swap, insert, combine, and redistribute proportions that are invested in assets. Besides that, there is a local improvement that combines the proportion of two consecutive assets.

The results of the MO-BNVS are compared with those of the quadratic bi-objective mean-variance model solved with an approach based on the ϵ-constraint method. The MO-BVNS could obtain Pareto fronts with a large number of non-dominated points, although several points seemed to be dominated by those of the Pareto front of the MV model. One advantage of the MO-BVNS is the computational time that was required to construct the Pareto fronts, which is by far much smaller than that spent when solving the bi-objective MV model.

Future works are going to use the NGSA-II to compare with the proposed heuristic. In this case, different metrics are going to be used to assess the performance of each method in terms of convergence and diversity among points of the optimal Pareto set. Another direction is related to investigate other neighborhood structures and a local search based on the variable neighborhood descent method.

Acknowledgments. The authors would like to thank Intel, the National Counsel of Technological and Scientific Development (CNPq - grants 202006/2018-2 and 308312/2016-3), the State of Goiás Research Foundation (FAPEG), and the State of São Paulo Research Foundation (FAPESP - grant 2013/07375-0) for their financial support.

References

1. Anagnostopoulos, K.P., Mamanis, G.: A portfolio optimization model with three objectives and discrete variables. Comput. Oper. Res. **37**(7), 1285–1297 (2010)
2. Anagnostopoulos, K.P., Mamanis, G.: The mean–variance cardinality constrained portfolio optimization problem: an experimental evaluation of five multiobjective evolutionary algorithms. Expert Syst. Appl. **38**(11), 14208–14217 (2011)
3. Chang, T.J., Meade, N., Beasley, J., Sharaiha, Y.: Heuristics for cardinality constrained portfolio optimisation. Comput. Oper. Res. **27**(13), 1271–1302 (2000)
4. Chen, C., Zhou, Y.: Robust multiobjective portfolio with higher moments. Expert Syst. Appl. **100**, 165–181 (2018)
5. Corne, D.W., Knowles, J.D., Oates, M.J.: The pareto envelope-based selection algorithm for multiobjective optimization. In: Schoenauer, M., et al. (eds.) PPSN 2000. LNCS, vol. 1917, pp. 839–848. Springer, Heidelberg (2000). https://doi.org/10.1007/3-540-45356-3_82
6. De, M., Mangaraj, B.K., Das, K.B.: A fuzzy goal programming model in portfolio selection under competitive-cum-compensatory decision strategies. Appl. Soft Comput. **73**, 635–646 (2018)
7. Deb, K., Pratap, A., Agarwal, S., Meyarivan, T.: A fast and elitist multiobjective genetic algorithm: NSGA-II. IEEE Trans. Evol. Comput. **6**(2), 182–197 (2002)

8. Duarte, A., Pantrigo, J.J., Pardo, E.G., Mladenovic, N.: Multi-objective variable neighborhood search: an application to combinatorial optimization problems. J. Glob. Optim. **63**(3), 515–536 (2014). https://doi.org/10.1007/s10898-014-0213-z

9. Fieldsend, J.E., Matatko, J., Peng, M.: Cardinality constrained portfolio optimisation. In: Yang, Z.R., Yin, H., Everson, R.M. (eds.) IDEAL 2004. LNCS, vol. 3177, pp. 788–793. Springer, Heidelberg (2004). https://doi.org/10.1007/978-3-540-28651-6_117

10. Hansen, P., Mladenović, N., Moreno Pérez, J.A.: Variable neighbourhood search: methods and applications. 4OR **6**(4), 319–360 (2008)

11. Kalayci, C.B., Ertenlice, O., Akbay, M.A.: A comprehensive review of deterministic models and applications for mean-variance portfolio optimization. Expert Syst. Appl. **125**, 345–368 (2019)

12. Konak, A., Coit, D.W., Smith, A.E.: Multi-objective optimization using genetic algorithms: a tutorial. Reliab. Eng. Syst. Saf. **91**(9), 992–1007 (2006). Special Issue - Genetic Algorithms and Reliability

13. Konno, H., Yamazaki, H.: Mean-absolute deviation portfolio optimization model and its applications to tokyo stock market. Manage. Sci. **37**(5), 519–531 (1991)

14. Kumar, D., Mishra, K.: Portfolio optimization using novel co-variance guided artificial bee colony algorithm. Swarm Evol. Comput. **33**, 119–130 (2017)

15. Liagkouras, K., Metaxiotis, K.: A new probe guided mutation operator and its application for solving the cardinality constrained portfolio optimization problem. Expert Syst. Appl. **41**(14), 6274–6290 (2014)

16. Macedo, L.L., Godinho, P., Alves, M.J.: Mean-semivariance portfolio optimization with multiobjective evolutionary algorithms and technical analysis rules. Expert Syst. Appl. **79**, 33–43 (2017)

17. Markowitz, H.: Portfolio selection. J. Financ. **7**(1), 77–91 (1952)

18. Markowitz, H.: Portfolio Selection: Efficient Diversfication of Investments, vol. 7. Wiley, New York (1959)

19. Mladenović, N., Hansen, P.: Variable neighborhood search. Comput. Oper. Res. **24**(11), 1097–1100 (1997)

20. Nocedal, J., Wright, S.: Numerical Optimization. Springer, New York (2006). https://doi.org/10.1007/978-3-540-35447-5. https://www.springer.com/br/book/9780387303031

21. Ong, C.S., Huang, J.J., Tzeng, G.H.: A novel hybrid model for portfolio selection. Appl. Math. Comput. **169**(2), 1195–1210 (2005)

22. Panadero, J., Doering, J., Kizys, R., Juan, A.A., Fito, A.: A variable neighborhood search simheuristic for project portfolio selection under uncertainty. J. Heuristics 1–23 (2018). https://doi.org/10.1007/s10732-018-9367-z

23. Schaffer, J.D.: Multiple objective optimization with vector evaluated genetic algorithms. In: Proceedings of the 1st International Conference on Genetic Algorithms, pp. 93–100. L. Erlbaum Associates Inc., Hillsdale (1985)

24. Skolpadungket, P., Dahal, K., Harnpornchai, N.: Portfolio optimization using multi-objective genetic algorithms. In: 2007 IEEE Congress on Evolutionary Computation, pp. 516–523, September 2007

25. Yitzhaki, S.: Stochastic dominance, mean variance, and Gini's mean difference. Am. Econ. Rev. **72**(1), 178–185 (1982)

26. Zitzler, E., Laumanns, M., Thiele, L.: SPEA2: improving the strength pareto evolutionary algorithm. Technical report, Computer Engineering and Networks Laboratory, Swiss Federation of Technology, Zurich (2001)

Local Search Approach for the $(r|p)$-Centroid Problem Under ℓ_1 Metric

Ivan Davydov[1,2](✉) and Petr Gusev[2]

[1] Sobolev Institute of Mathematics, Novosibirsk, Russia
vann.davydov@gmail.com
[2] Department of Mechanics and Mathematics, Novosibirsk State University,
Novosibirsk, Russia
pgusev@outlook.com
https://www.researchgate.net/profile/Ivan_Davydov

Abstract. In the $(r \mid p)$-centroid problem, two players, called the Leader and the Follower, open facilities to service customers. We assume that customers are identified with their location on the plane, and facilities can be opened anywhere on the plane. The Leader opens p facilities. Later on, the Follower opens r facilities. Each customer patronizes the closest facility. The distances are calculated according to ℓ_1-metric. The goal is to find the location of the Leader's facilities maximizing her market share. We provide the results on the computational complexity of this problem and develop a local search heuristic, based on the VNS framework. Computational experiments on the randomly generated test instances show that the proposed approach performs well.

Keywords: Variable neighborhood search · Stackelberg game · (r | p)-centroid · Facility location · Bilevel programming · Manhattan metric

1 Introduction

This paper addresses a well-known Stackelberg facility location game on a two–dimensional plane, namely the $(r|p)$–centroid problem. The problem can be stated as follows. We are given a set of customers that are concentrated at a finite number of points in the two–dimensional plane. We assume that a weight w_i, which represents the demand, is assigned to each customer. At the first stage of the game, a player, called the Leader, opens p facilities. At the second stage, another player, called the Follower, opens r facilities. Both players are able to open their facilities anywhere on the plane. At the third stage, each customer chooses the closest opened facility as a supplier. We consider the case when the distances are defined according to ℓ_1-metric. In case of ties, the Leader's facility is preferred. Customer's service induce an income of w_i to the supplier.

Supported by Russian Science Foundation (project no. 17-11-01021).

R. Benmansour et al. (Eds.): ICVNS 2019, LNCS 12010, pp. 81–94, 2020.
https://doi.org/10.1007/978-3-030-44932-2_6

Each player aims to maximize its own market share. The goal of the game is to find p points for the Leader facilities to maximize her market share.

The $(r \mid p)$–centroid problem was first studied by Hakimi [8]. He explored the formulation considering location on a network, reviewed its applications and variations and showed that the problem is NP–hard. Since then, decent amount of publications has been dedicated to the centroid problem. In study [10], by Kress and Pesch, the authors presented an overview of the latest developments in the area of sequential competitive location problems. Special attention was paid to problems defined on networks. Authors also provide a highlight of work that has been done in the field of $(r \mid X_p)$–medianoid and $(r \mid p)$–centroid problems. The problems in the review were studied under different formulations, metrics and location spaces. Bauer et al. in [2] explored the $(1—1)$-centroid problem on the network with different types of clients behaviour. The optimal location for the Leader in this work is defined as location, such that no point on the network with higher expected value exists. Authors showed that the set of such optimal points consists only of network vertices assuming that clients are located only at the vertices too. As a main result authors proposed two algorithms which can determine all optimal points of a network in polynomial time. For the case $p = r = 1$ on the Euclidean plane an exact polynomial-time algorithm is introduced in [7]. The author addressed two problems in his work, one of which is how to locate new facility in order to gain more value from the existing one. A more general case with an arbitrary values of p and r is considered in [3]. Authors presented an alternating heuristic for the $(r \mid p)$–centroid problem on two-dimensional Euclidean plane combined with a greedy heuristic for the Follower's problem. The development of this approach can be found in [9]. An exact method is applied for the Follower problem. In order to improve the results of the Leader's problem a clustering procedure is introduced combined with exact polynomial algorithm for $(1 \mid 1)$–centroid problem. More in-depth review of problem's complexity and properties can be found in [1]. Ashtiani summarizes recent publications devoted to competitive facility location problems and introduces the classification based on specific components, for example, variables, competition type, solution space, etc. Based on this taxonomy a comparison of various studies is provided.

A gap we intend to fill in this work is devoted to ℓ_1-metric case of this problem. Although, this case has various applications and is of certain theoretical interest, to our knowledge no papers have considered such formulations before. We present a mathematical model for the $(r|p)$-centroid problem under ℓ_1 metric and propose a local search heuristic combined with an exact approach for the Follower problem. We consider the $(r \mid X_{p-1} + 1)$-centroid sub-problem where the Leader moves exactly one facility, searching for its optimal relocation. We use this problem in order to find the best neighboring solution in the Swap neighborhood. To reduce the computational efforts we use the concept of randomized neighborhoods and apply a local descent algorithm to evaluate the goal function during the neighborhood exploration. We also adopt a maximum clique approach from [13] to reduce the computational complexity of the Follower's problem.

The paper is organized as follows. In Sect. 2 we present a comprehensive formulation and provide a mathematical model for the problem. In Sect. 3 we propose an exact approach to tackle the Follower's problem. Section 4 provides the formulation for the auxiliary problem of the Leader, called the $(r \mid X_{p-1} + 1)$-centroid problem. Sect. 5 provides results of computational experiment and Sect. 6 concludes the study.

2 Mathematical Model

First let us provide a mathematical model for the $(r \mid p)$–centroid problem in a following formulation: let n define the number of *clients* located on a two–dimensional plane. Each client j is associated with the positive weight w_j representing the demand. The set X of size p represents points, in which the *Leader* located his facilities. Similarly, Y defines the set of *Follower's* r points, in which his facilities are located. The ℓ_1 distance from client j to the closest facility patronized by the *Leader* is defined as $d(j, X)$. The ℓ_1 distance to the closest *Follower's* facility is denoted by $d(j, Y)$. Customer j chooses *Follower's* facility instead of *Leader's* one if $d(j, Y) < d(j, X)$ and chooses *Leader's* facility otherwise. By

$$U(Y \prec X) := \{j \mid d(j, Y) < d(j, X)\}$$

let us denote the set of customers, who prefer facility from Y instead of X. The *Follower's* profit, gained by locating his facilities at Y, in response to the *Leader's* facilities in X is represented by:

$$W(Y \prec X) := \sum (w_j \mid j \in U(Y \prec X)).$$

In response to the *Leader's* solution X, the *Follower* tries to maximize his gain. The maximal value $W^*(X)$ is defined to be

$$W^*(X) := \max_{Y, |Y| = r} W(Y \prec X).$$

Later on, we will refer to this problem as a *Follower's problem*. The whole market is divided between competitors, so the *Leader* tries to minimize the market share of the *Follower*. This minimal value $W^*(X^*)$ is defined as

$$W^*(X^*) := \min_{X, |X| = p} W^*(X).$$

For the best solution X^* of the Leader, her market share is $\sum_{j=1}^{n} w_j - W^*(X^*)$.

In the $(r \mid p)$–*centroid problem* the aim is to find X^* and $W^*(X^*)$. We claim that the $(r \mid p)$-centroid problem under ℓ_1 metric is Σ_2^p-hard, while the Follower's problem is NP–hard.

3   Follower's Problem

In this section we present an exact approach for the *Follower*'s problem. It can be formulated as an *ILP* and solved using the branch and bound method. The exact algorithm for the *Follower*'s problem consists of two stages. At the first stage the problem is discretized. This can be done in the following way. In order to "capture" customer j, a *Follower* needs to place his facility closer than the nearest *Leader*'s facility, i.e. at a distance less than $d(j, X)$ from client j. For each client j we associate a disk D_j of radius $d(j, X)$ with the center at the point where the client j is located. If two or more disks form the intersection, then by placing the facility inside of the intersection the *Follower* captures more than one customer at once. Disks and their intersections divide the plane into a number of areas. For each area we can calculate the income that the *Follower* will receive by placing his facility in this area.

Basic concepts of the *Follower*'s problem in terms of ℓ_1 metric can be defined as follows. A circle is a set of points with a fixed distance, called the radius, from a point called the center. In ℓ_1 metric, distance is determined by a different metric than in Euclidean geometry and the shape of the circles changes as well. Under ℓ_1 norm the circles have a square shapes with sides oriented at a 45° angle to the coordinate axes. For a fixed solution X of the *Leader*, for each client j, we introduce a square D_j with radius $d(j, X)$ centered in client location point. If the *Follower* would place the facility in region D_j he would for sure capture the client j. Let us consider the resulting intersection of each set of two or more such squares, which will be called a *region*. In order to simplify the computation of the regions described above, we used the following result for R^2:

Lemma 1. *Let $P_1, ..., P_n \subset R^2$ be rectangles with sides parallel to the coordinate axes, such that every two rectangles intersect. Then all rectangles have a nonempty intersection.*

Curious reader can find the proof of this lemma in [12].

This lemma helps to reduce computation time in the algorithm of searching for interesting points for facility location when solving the *Follower*'s problem. Instead of computing an intersection of each set of squares we can now count only pairwise intersections in order to find out whether these rectangles have a point in common or not. In order to reduce computational time it is also important to find the regions which correspond to intersections that are maximal in inclusion. We propose to solve this problem by reformulating it in graph theory terms and solve it as *maximal clique problem*. To this end let us introduce the graph $G = (V, E)$, where V represents the set of clients and the edges E are calculated in the following way:

$$e_{i,j} \begin{cases} \in E \text{ if } D_i \text{ has not empty intersection with } D_j, \\ \notin E \text{ otherwise} \end{cases}$$

Therefore, we can think of finding the regions with maximal intersection as of *Maximal Clique problem*. A comprehensive review of publications on this well-studied topic can be found in [14]. In order to solve this problem we have applied the approach proposed by Tanakaa et al. [13] with the notable complexity $(O(3^{n/3}))$. Here we briefly describe the scheme of the algorithm.

Definition 1. *For a vertex $(v \in V)$, let $(\Gamma(v))$ be the set of all vertices that are adjacent to (v) in $G = (V, E)$.*

The scheme of the algorithm is presented below:

```
procedure CLIQUES(G) /* Graph G=(V,E) */
begin
0: /* Q := ∅ */
1: EXPAND(V, V)
end of CLIQUES

procedure EXPAND(SUBG, CAND)
begin
2: if SUBG = ∅
3:    then maximal clique is found.
  return
4: else u := a vertex u in SUBG that maximizes | CAND ∩ Γ(u) |;
5:    while CAND - Γ(u) ≠ ∅
6:        do q := a vertex in (CAND - Γ(u));
7:           print(q, ",");
8:           SUBG_q := SUBG ∩ Γ(q);
9:           CAND_q := CAND ∩ Γ(q);
10:          EXPAND(SUBGq, CANDq);
11:          CAND := CAND - q;
12:          /* one step out of the reqursion */
   od
     fi
end of EXPAND
```

The general procedure expands the desired clique until the maximal one is found. The EXPAND procedure starts from an empty set and expands a global variable Q step by step by applying a recursive procedure EXPAND to the set of vertices V and its succeeding induced subgraphs to search for larger and larger complete subgraphs until they reach maximal ones. Here SUBG is a full set of candidates to be added to the clique. SUBG consists of all vertices from V that are adjacent to all vertices of Q. Let FINI denote a subset of vertices of SUBG that have already been processed by the algorithm. The set of remaining candidates for expansion is denoted by CAND: CAND = SUBG?FINI. $\Gamma(u)$ denotes the set of al vertices adjacent to u.

If the *Follower* would place his facility inside a region, he captures all the clients, whose squares D_j form the following region. The resulting number of regions may be large, however, we can choose the regions formed by maximal intersections and eliminate the dominated ones. The algorithm presented above allows to obtain these regions effectively. As we need to solve the Follower's problem many times during the solution of the Leader's problem, it is extremely important to reduce the computational costs of the solution procedure.

We can indicate the clients who patronize *Follower*'s facility if he would open one inside the region k by introducing a matrix a, such that $a_{k,j} = 1$ if the facility of the *Follower* in region $k \in K$ captures client j and $a_{k,j} = 0$ otherwise . We formulate the *Follower*'s problem in form of *ILP* by introducing two sets of the decision variables:

$$y_k = \begin{cases} 1 & \text{if the Follower open his facility inside region } k \in K \\ 0 & \text{otherwise} \end{cases}$$

$$z_j = \begin{cases} 1 & \text{if the Follower captures client } j \\ 0 & \text{otherwise} \end{cases}$$

The *Follower's* problem now can be formulated as the maximum capture problem:

$$\max \sum_{j=1}^{n} w_j z_j \tag{1}$$

subject to

$$z_j \le \sum_{k=1}^{K} a_{kj} y_k \quad j = 1, ..., n \tag{2}$$

$$\sum_{k=1}^{K} y_k = r, \quad y_k, z_j \in \{0, 1\}. \tag{3}$$

The objective function targets the maximization of *Follower*'s gain. First constraint guarantees that client j can patronize the Follower's facility only if there is one, opened in a proper region. Second constraint enforces the Follower to open exactly r facilities.

With a reduced number of regions K this problem can be easily solved to optimality with the help of any ILP solver.

4   Local Search Algorithm

At the first step we address the *Follower's* problem (lower level problem). In this case the location of *Leader*'s facilities is known and given as an input. In order to tackle the upper level problem (the *Leader's* one) we propose a VNS-based approach [11]. To explore the neighborhood we consider the $(r|X_{p-1} + 1) - centroid$ problem where the Leader has a set of $p - 1$ facilities and wants to open another facility in the best position [9]. We applied an alternating heuristic from [4] as an alternative approach to achieve baseline results. The idea of this heuristic is as follows. For a given *Leader*'s solution X, in response, *Follower* computes his best possible solution Y. Next, the *Leader* may decide to move his facilities in order to improve his solution. Among other options, one is to use the solution Y of the *Follower* to locate the facilities of the *Leader*. Thus, on this step the *Leader* acts as *Follower*, trying to find her best response to the *Follower*'s move. This process repeats until the stopping criterion is met. The scheme of the method is as follows:

1. Create a starting solution X for the Leader
while *not termination condition* **do**
    2.1 Find the best solution Y for the Follower against the solution X;
    2.2 Find the best solution X for the Leader against the solution Y;
end

Algorithm 1: Alternating heuristic

The algorithm begins with random initial solution. The stopping criterion is the number of iterations performed. As a result of the *Follower's* problem solution we do not have precise location coordinates, but only the regions where the Follower locate his facilities. Therefore the procedure of finding the exact coordinates for the facilities must be used in order to proceed to the alternating step. All points inside each region are equivalent from the Followers point of view. In contrast, it is not true in the *Leader's* case. At Step 2.2 of the alternating process the facilities of the *Leader* have to be opened in exact location. To reduce the computational costs of the iterative process, we take the center points of the regions as locations.

Consider the $(r|X_{p-1} + 1)$-centroid problem, where the Leader has $p - 1$ open facilities and intends to open one more in the best position. Each candidate location for new facility can be evaluated by solving the Follower's problem. The point maximizing the *Leader's* income is the desired one. As in the *Follower's* problem, we associate with each client j a disk D_j of radius $R_j = d(j, X_p - 1)$. Note that the disks divide the plane into a fixed number of regions. When the new facility is opened by *Leader*, some of the disks and the corresponding regions may shrink or even disappear as the radius of the disks cannot increase. Note that all points within the same region are of equal value for the *Follower*.

According to the Helly theorem, for $d = 2$ intersections of pairs disks define the way all disks intersect. Number of regions change only if the structure of intersection changed at least in one intersecting pair.

Let D_1 and D_2 be disks with radius R_1 and R_2 respectively. Let's consider all possible locations of new Leader's facility at which at least one region disappears:

1. Disks do not intersect. If leader opens his facility in point j_1 or j_2, corresponding disk disappears, which means that now we have new input data for the Follower's problem. All the other points are equivalent and make no interest for the Leader.
2. Disks do intersect, and center of one, say j_1 is inside D_2. In this case, we have two sub-regions of interest and if Leader would place the facility inside a union of those, the intersection will disappear: *Rectangle, formed by diagonals* - is the rectangle, formed by intersection of corresponding diagonals of the disks; *Sub-disk* - is basically a rhombus with center in point j_2. Its radius equals the closest ℓ_1 distance to another disk.

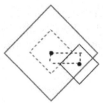

Fig. 1. Center of one disk is inside of other

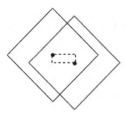

Fig. 2. Centers are inside the area of opposite disk

3. Disks do intersect and their centers are inside each other. In this case, leader only can eliminate the intersection by placing the facility inside *Rectangle formed by diagonals* of that intersection.
4. Disks do intersect, neither of centers is inside another disk. In this case we have region of interest formed by following sub-regions: Sub-disk of D_1 ∪ Sub-disk of D_2 ∪ *Rectangle, formed by diagonals.*

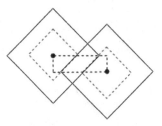

Fig. 3. Regular intersection

Lets consider now all clients at once. The above cases describe all points, segments and discs that make up the boundaries of regions. Regardless of the choice of a point inside the region, by placing a new facility at this point, the Leader creates the same instance of the Follower problem and therefore gets the same goal function value. The Follower problem can change only when intersections vary. As mentioned above, in the case of equal distances to the facilities of the Leader and the Follower, the client prefers the Leader. This means that any point on the boundary of the region is not worse than internal. Thus, it suffices to consider only points of intersection of the region boundaries and calculate the value of the goal function for each of them. There are no more than $O(n^2)$ segments and disks, and therefore no more than $O(n^4)$ intersection points.

To solve the problem we propose a heuristic based on a variable neighborhood search (VNS). There were mainly two reasons to make that choice. The first one is that VNS approach in know to be very efficient when solving the location problems [6]. The second reason is that the neighborhood structure that can be obtained from the results above perfectly correlates with the scheme of the approach. We use the Swap (k, l) neighborhood with different values of k and l. In this neighborhood, k facilities of the *Leader* moves to new places, but not further than at a distance l from their current position. At the diversification step, the values $l_i = 50i$, $i = 2, ..., i_m ax$ and $k = 1, ..., k_m ax$ are used. At the local improvement step $l_1 = 50$ and $k = 1$. As the number of candidate points for relocation of the facility of the Leader is large enough, during the local search step we consider only the closest vicinity of current location of the transferred facility. During the shake step the area to consider becomes larger and larger with the growth of i parameter.

Below we present the scheme of the variable neighborhood search for the $(r|p)$–centroid problem:

1. Initialization. Generate the initial Leader's solution X and calculate its gain $F(X)$; determine $i_m ax$, $k_m ax$, and stopping criterion.
while *not termination condition* **do**
$\quad$ $i = 1, k = 1$
$\quad$ **while** $i \le i_{max}$ *and* $k \le k_{max}$ **do**
$\quad\quad$ 2.1 Shake (Diversification). Generate the solution X' by selecting it from the (k, l_i)–Swap neighborhood arbitrarily;
$\quad\quad$ 2.2 Local Search. Apply the local improvement method, taking X' as the initial solution; denote by X'' the obtained local optima;
$\quad\quad$ 2.3 Move or not. If $F(X) < F(X'')$, then $X = X''$, $i = 1, k = 1$. Else $i = i + 1$
$\quad$ **end**
end

Algorithm 2: VNS scheme

As the stopping criterion, the running time of the algorithm is used. The initial solution is generated using alternating heuristic. The most computationally intensive part of the VNS scheme introduced above is Step 2.2. We propose two approaches in order to decrease the computation time. One of them is to divide the neighborhood (k, l)–Swap into several disjoint neighborhoods. These neighborhoods are investigated sequentially using the first improvement rule for local search in step 2.2 of the algorithm. The second idea is to use probabilistic neighborhoods. Instead of scanning all elements of a neighborhood in a row, a probabilistic $(k, l)_q$ neighborhood is used. Each element from the (k, l)–Swap neighborhood is included into $(k, l)_q$ neighborhood with a given probability q.

This approach can significantly reduce the complexity of the calculations without losing the quality of the solutions obtained [5].

5   Computational Experiments

The presented algorithm was implemented in the *Python3* environment and was tested on examples from the *Discrete Placement Electronic Library* [15]. In all examples, $n = 100$ and clients are placed with a uniform distribution on the square with side length 7000. Two types of demand (purchasing power) are considered: equal, $w_j = 1$, and differentiated, when purchasing power is chosen with a uniform distribution from the interval, $w_j \in [1, 200]$. For all instances, the behavior of the algorithm is studied for $p = r = 10$.

The experiments were carried out on a PC with Intel? Core? i7 4700HQ processor and 8 GB RAM. The termination criterion for alternating heuristic was set as 50 consecutive iterations without improvement. The termination criterion for VNS was set as 10 min of computational time. These values were obtained during the test process. It was established that, there is no sense to conduct the computations any longer as no improvement of the solution occurs.

Table 1. Alternating heuristic

No.	Initial	Improvement %	Total (%)	Time (s)
111	625	+25.58%	2848 (32.78%)	38
211	376	+29.33%	3462 (32.91%)	38
311	421	+26.35%	2885 (30.85%)	42
411	803	+22.43%	3030 (30.52%)	44
511	898	+21.80%	3131 (30.57%)	39
611	734	+23.42%	2984 (31.35%)	39
711	649	+27.13%	3688 (32.93%)	37
811	1200	+17.59%	2881 (30.15%)	41
911	687	+32.60%	3389 (32.6%)	48
1011	946	+30.84%	3154 (30.84%)	51

Tables 1 and 2 shows the results of experiments for 10 instances, the purchasing power of customers $w_j \in [1, 200]$, both for alternating heuristic and VNS approach. For Table 1, first column represents the benchmark number. The second column shows the initial market share of the *Leader* for a random initial solution X. The third column represents market share improvement percentage, obtained by alternating heuristic. Total Leader's gain is shown in 4-th column of the Table 1.

Table 2 is devoted to the local search results. Solution gain, obtained by alternating heuristic is shown in second column. Third column represents the

Table 2. Alternating heuristic and local search

No.	Alt. h.	Improvement %	Total (%)	Time (s)
111	2848	+3.71%	3171 (36.49%)	113
211	3462	+2.58%	3734 (35.49%)	236
311	2885	+1.5%	2994 (32.02%)	119
411	3030	+2.69%	3297 (33.21%)	381
511	3131	+1.45%	3280 (32.02%)	136
611	2984	+1.25%	3103 (32.60%)	113
711	3688	+1.63%	3871 (34.56%)	283
811	2881	+2.98%	3166 (33.13%)	261
911	3389	+2.22%	3620 (34.82%)	167
1011	3154	+3.97%	3560 (34.81%)	433

improvement of the VNS regarding alternating heuristic result. Forth column contains the total market share of the Leader. Fifth column in both tables represents the running time until the best known solution is found. From these tables we observe that results, obtained by alternating heuristic could be improved by 2 to 4% with the help of our approach. The resulting market share of the leader is about one third of the market. Comparing these results with the corresponding goal function values for the same instances under different metrics, we observe the following. In the discrete case, when facilities can be opened only in a fixed number of points, the Leader's share is about a half (48–51%) of the market. In a continuous case, under Euclidean metric, the share of the Leader become 10% lower, and is about 39–41%. Apparently the ell_1 case appears to be the most unprofitable for the Leader. The reason might be that the distances in ℓ_1 are usually significantly larger, thus the disks are larger too, giving the Follower more opportunities to capture another customer.

In order to verify the results we have conducted two additional experiments aiming to improve the solutions obtained. We considered the instance 111. During the experimental runs we set the termination criteria for the VNS-approach to 84000 seconds of computational time (one day). During the first run we used the same neighborhood structure as before. During the second run, instead of using the intersections of discs as candidate points, we used the grid vertices to relocate Leader's facilities. Grid step was set to 7 (while). Regardless the neighborhood structure we were unable to improve the results of a "short" run, although we faced 12 different solutions with the same goal function value (of 3171) during the search. This observation allows us to assume that the solutions found are probably not far from being optimal. The construction of effective upper bounds on the Leader's objective function value seems to be an extremely interesting direction for further research, as these bounds will help us to approve our brave assumption.

Figure 4 shows the landscape of the goal function. We consider the instance 111 from the benchmark library. We generated the solution with the algorithm presented. Then, we selected one facility and move it along the grid with step 20. For each movement of that facility we solved the *Follower*'s problem. Then, for each point, the resulted *Follower*'s gain was subtracted from the total market value and assigned to that point as *Leader*'s gain, if he would place the facility in that point. At the end of this procedure we had 122,500 values, which we used as the height in our graphs, in which the greater the height represent the greater *Leader*'s income.

There are few peaks with different values of the goal function. In Fig. 5a the same landscape is presented from other point of view. Finally, in Fig. 5b the top view of the same landscape is presented. There are several rectangles, triangles and their intersections. The upper-right region is considered the most attractive for the Leader. These pictures graphically proves the effectiveness of the proposed local search approach. Indeed all the structures, listed above can be seen from the top view.

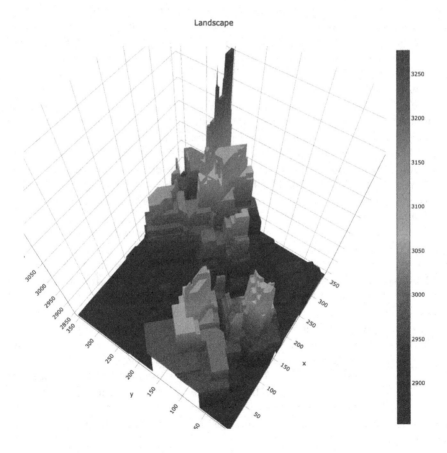

Fig. 4. Landscape side view

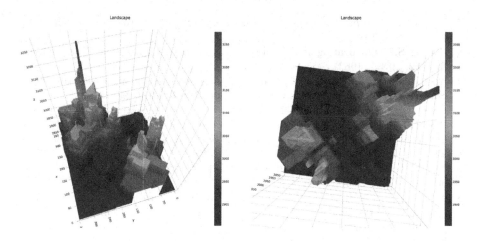

Fig. 5. a. Landscape side view, b. Landscape top view

6   Conclusion

In this study we considered the $(r|p)$-centroid problem on a plane. We provide an overview of studies devoted to the topic of research. The main result of this paper is the implementation of the algorithm for solving the $(r|p)$–centroid problem in ℓ_1 metric. An alternating heuristic and a variable neighborhood search approach were developed. We presented a computational experiments which shows that the proposed algorithm preforms well. The result were compared to those achieved in works of other authors in different metrics. The landscape view of the $(r|X_{p-1}+1)$–centroid problem solution confirms our claim on how to accomplish a search for good *Leader* facility location points.

References

1. Ashtiani, M.: Competitive location: a state-of-art review. Int. J. Ind. Eng. Comput. **7**(1), 1–18 (2016)
2. Bauer, A., Domshke, W., Pesch, E.: Competitive location on a network. Eur. J. Oper. Res. **66**, 372–391 (1993)
3. Bhadury, J., Eiselt, H.A., Jaramillo, J.H.: An alternating heuristic for medianoid and centroid problems in the plane. Comput. Oper. Res. **30**, 553–565 (2003)
4. Carrizosa, E., Davydov, I., Kochetov, Y.: A new alternating heuristic for the $(r|p)$-centroid problem on the plane. In: Klatte, D., Lüthi, H.J., Schmedders, K. (eds.) Operations Research Proceedings, pp. 275–280. Springer, Heidelberg (2011). https://doi.org/10.1007/978-3-642-29210-1_44
5. Davydov, I., Kochetov, Y.: VNS-based heuristic with an exponential neighborhood for the server load balancing problem. Electron. Notes Discrete Math. **47**, 53–60 (2015)
6. Davydov, I., Kochetov, Yu., Mladenovic, N., Urosevic, D.: Fast metaheuristics for the discrete (r—p)-centroid problem. Autom. Remote Control **75**, 677–687 (2014)

7. Drezner, Z.: Competitive location strategies for two facilities. Reg. Sci. Urban Econ. **12**, 485–493 (1982)
8. Hakimi, S.L.: On locating new facilities in a competitive environment. Eur. J. Oper. Res. **12**, 29–35 (1983)
9. Kochetov, Y., Davydov, I., Carrizosa, E.: A local search heuristic for the $(r|p)$-centroid problem in the plane. Comput. Oper. Res. **52**, 334–340 (2014)
10. Kress, D., Pesch, E.: Sequential competitive location on networks. Eur. J. Oper. Res. **217**, 483–499 (2012)
11. Mladenović, N., Hansen, P.: Variable neighborhood search. Comput. Oper. Res. **24**, 1097–1100 (1997)
12. Pak, I.: Lectures on discrete and polyhedral geometry (2008). http://www.math.ucla.edu/~pak/geompol8.pdf
13. Tomitaa, E., Tanakaa, A., Takahashia, H.: The worst-case time complexity for generating all maximal cliques and computational experiments. Theoret. Comput. Sci. **363**, 28–42 (2006)
14. Wu, Q., Hao, J.: A review on algorithms for maximum clique problems. Eur. J. Oper. Res. **242**(3), 693–709 (2015)
15. Discrete Location Problems. Benchmark library. http://math.nsc.ru/AP/benchmarks/index.html

Optimization of Maintenance Planning and Routing Problems

Lamiaa Dahite[1,2]([envelope]), Abdeslam Kadrani[1], Rachid Benmansour[1,3],
Rym Nesrine Guibadj[2], and Cyril Fonlupt[2]

[1] SI2M, Institut National de Statistique et d'Economie Appliquée, Rabat, Morocco
lamiaadahite@gmail.com, {akadrani,r.benmansour}@insea.ac.ma
[2] LISIC, EA 4491, Université du Littoral Côte d'Opale, Calais, France
{rym.guibadj,cyril.fonlupt}@univ-littoral.fr
[3] LAMIH UMR CNRS 8201, Université Polytechnique hauts de France,
Valenciennes, France

Abstract. This paper addresses a problem that maintenance service providers often face: determining the best routing-maintenance policy for all technicians and machines. It consists of defining for each technician the sequence of the maintenance operations to perform so that the total expected costs are minimized while maintaining a high service level on machines availability. We propose in this paper a mathematical model with different objective functions which integrates both routing and maintenance considerations. To solve the problem, we propose constructive and improvement heuristics and a Variable Neighborhood Search that uses sequentially different neighborhood structures. The performance of our algorithms is evaluated using new generated instances. Results provide strong evidence of the effectiveness of our heuristic approach.

Keywords: Time based maintenance · Random failures · Vehicle routing problem · Variable Neighborhood Search · Variable Neighborhood Descent · Heuristics

1 Introduction

Maintenance is a primary service in industry, especially when failures cause important damages on personnel and environmental safety. The companies mostly outsource their maintenance operations to a service provider in order to focus on their core business. The maintenance provider agrees to satisfy the requirements of its customers by ensuring good quality maintenance services at the lowest overall cost. Finding when and how to execute the maintenance are therefore major concerns. In this paper, we propose a mathematical model and neighborhood search approaches to tackle the problem in the case of time based preventive maintenance. The proposed model aims to jointly integrate maintenance and routing considerations in the technicians assignment to the

© Springer Nature Switzerland AG 2020
R. Benmansour et al. (Eds.): ICVNS 2019, LNCS 12010, pp. 95–111, 2020.
https://doi.org/10.1007/978-3-030-44932-2_7

maintenance tasks. Three objective functions are tested. For each objective, a General Variable Neighborhood Search, a Variable Neighborhood Descent and a Best Improvement Local Search are tested. We have designed greedy constructive heuristics to find an initial solution. Our algorithms are therefore compound integrating construction and improvement phases. The remainder of this paper is organized as follows: Sect. 2 gives an overview of the related literature. Section 3 describes the problem and provides its mathematical formulation. A description of the proposed solution approaches and details on their implementation are shown in Sect. 4, followed by experimental results in Sect. 5. Finally, concluding remarks and directions for future research are presented in Sect. 6.

2   Literature Review

Most of the papers that deal with routing and maintenance optimization consider separated routing and maintenance models. Indeed, we distinguish two main streams of research in the literature: the first one use the routing model to plan maintenance operations and the second combines maintenance and routing models.

The main problems defined in the first category are: the Technician Routing and Scheduling Problem (TRSP) [3,5], workforce scheduling, the Technician and Task Scheduling Problem (TTSP) [1], the Service Technician Routing and Scheduling Problem (STRSP) [2] and Geographically Distributed Asset Maintenance Problems (GDMP) [8]. Cordeau et al. [1] propose a construction heuristic and an adaptive large neighborhood search heuristic to solve the technician and task scheduling problem arising in a large telecommunications company. Each technician is specialized in different tasks with different skill levels and can perform tasks requiring skills at lower levels than his own. In addition, each operation is associated to a specific time window and has skills requirements. The proposed heuristic defines technicians teams and assigns tasks to each team. Kovacs et al. [2] define the service technician routing and scheduling problem which considers routing and outsourcing costs, skills and team building. The problem is then solved using an adaptive large neighborhood search algorithm. Zamorano and Stolletz [3] present a mixed integer program for the multiperiod TRSP. Technicians skills are considered in the assignment to tasks and to teams of technicians. The skill constrained tasks have to be realized within a time window that can last several periods. The objective is minimizing total travel costs, waiting costs and overtime costs. A branch-and-price algorithm is then proposed to solve this problem. Mathlouthi et al. [4] propose a mixed integer programming model for the multi-attribute technician routing and scheduling problem taking into accounts several attributes: technicians skills, task priorities, multiple time windows, parts inventory, breaks and overtime. Pillac et al. [5] highlight an approach based on an adaptive neighborhood search algorithm, first used to compute an initial solution, then to re-optimize it when a new request arrives over time to solve the dynamic routing and scheduling technicians problem (DTRSP). Technicians with the necessary skills, tools and spares must be assigned to the tasks while minimizing the total cost. Technicians have the option of replenishing tools and spares at the depot at any time to handle more requests.

The second stream of research is relatively recent. It focuses on the combination of maintenance and routing characteristics. Lopez-Santana et al. [6] propose a combined maintenance and routing (CMR) model in two stages. The first step is used to determine an optimal maintenance policy by calculating for each machine the frequency of preventive operations and their time windows while minimizing the cost related to preventive and corrective operations. In the second step, the output data of the maintenance model is considered as the input data of the routing model which aims to determine the routes that each technician must follow. We then obtain the start time of each preventive operations. With this information, the maintenance model is again solved. The procedure repeats until meeting the stopping criterion. No heuristic approach was proposed to solve the problem. Jbili et al. [7] considers a new variant of the vehicle routing problem where vehicles are subject to random breakdowns. Preventive maintenance operations should be performed by replacing critical vehicles components when reaching selected customers. The proposed model takes into account the vehicle's reliability, maintenance costs, maintenance duration, transportation cost, and penalties corresponding to late arrivals. The objective is to simultaneously determine the optimal routing sequence and the optimal sequence of preventive actions on the vehicles. They then proposed a genetic algorithm to solve large instances. Chen et al. [8] consider gully pot maintenance as a risk-driven maintenance problem. They propose a multi-period VRP which takes into account the risk impact of gully pot failure, estimated using meteorological information, and its failure behaviour on each day.

3   Problem Definition and Formulation

We consider several geographically distributed machines in multiple customers sites that are subject to random failures that can lead to sudden breakdowns and production losses. Preventive maintenance operations are scheduled for each machine at regular time intervals to limit the risk of sudden failures. Each preventive operation must be performed in the time window associated with it. When a machine suddenly breaks down, the team of technicians must repair it by performing a corrective operation. After each preventive or corrective operation, the machines are considered in a state similar to that of new machines. Maintenance operations have a duration and are associated with maintenance costs. The duration of the corrective operation is greater than the duration of the preventive operation. Likewise, the cost of a corrective maintenance operation is greater than the cost of a preventive maintenance operation. Our proposed model takes into account several important characteristics derived from the reality of today's industries. Teams of technicians with the same skills are available to perform preventive and corrective maintenance operations. If a preventive operation cannot be performed within its time window due to the high workload of technicians, we allow that a team of technicians arrives to an operation

i after the closing of its time windows b_i with a penalty cost. The lower bound of the time window however remains a hard constraint. When arriving before a_i, the technicians team has to wait until a_i to start the preventive maintenance operation.

The maintenance model uses information from equipment degradation. The stochastic aspect is integrated into the failure costs objective function (25) by including the failure probability and the reliability of each equipment. This is particularly useful in industries where breakdowns are very dangerous and influence safety. We test three main objective functions, the classical travel cost (24), the failure cost (25), and finally the total maintenance cost (26) used in [6]. In (25), the cost of failure represents the probability that a machine will fail before the start of the maintenance operation multiplied by the cost of the corrective action. We consider that the random variable of time to failure for each machine follows a Weibull law whose parameters are β_i and σ_i but it is possible to use any law that results from the historical data of failure of each machine. We consider β_i superior to 1 to consider the third part of the bathtub curve, which represents the wearout life of the machine. The wearout is the phenomenon of accelerating the risk of failure over time.

3.1 Notation of the Joint Maintenance and Routing Problem

The following sets and parameters are used to define the problem throughout the paper.

Sets

- $\mathcal{M} = \{1, \ldots, l\}$: set of machines locations.
- $\mathcal{P}_m = \{p_1^m, \ldots, p_{n_m}^m\}$: the set of preventive maintenance operations associated to machine $m \in \mathcal{M}$.
- $\mathcal{P} = \cup_{m \in M} \mathcal{P}_m$: the set of all preventive maintenance operations.
- $\mathcal{O} = \{1, \ldots, n\}$: the indexes of preventive maintenance operations ($n = \sum_{m \in \mathcal{M}} n_m$).
- $\mathcal{V} = \{0, \ldots, n+1\}$: set of vertices (departure depot 0, indexes of maintenance operations, arrival depot $n + 1$)
- $\mathcal{V}^o = \{0, \ldots, n\}$: set of origins vertices
- $\mathcal{V}^d = \{1, \ldots, n + 1\}$: set of destinations vertices
- $\mathcal{K} = \{1, \ldots, K\}$: set of technicians teams
- $\mathcal{A} = \{(i, j), i \in \mathcal{V}^o, j \in \mathcal{V}^d, i \neq j\}$: set of arcs between vertices

Global Parameters

- H: planning horizon for technicians and maintenance operations.
- n: total number of preventive maintenance operations to do in the planning horizon.
- K: total number of technicians teams.
- r: total number of tours needed.
- B: big number greater than the maximum duration of the tours.

Maintenance Parameters

- Cpm_m: cost of preventive maintenance (PM) operation for machine m.
- Ccm_m: cost of corrective maintenance (CM) operation for machine m.
- Cw_m: waiting cost per unit time due to the time waited before the arrival of the technicians team to machine m for a corrective operation (the cost of production loss).
- Tpm_m: duration of PM operation for machine m.
- Tcm_m: duration of CM operation for machine m.
- $M_m(\delta_m)$: the expected time to failure for machine m assuming the failure occurs before the preventive maintenance period δ_m.
- W_m: expected waiting time before the beginning of CM operation in case of sudden failure of machine m.
- $CM_m(\delta_m)$: total cost per unit time incurred in one cycle for machine m when the preventive maintenance period is δ_m.
- T_m: random variable of the time to failure of machine m.
- β_m: the shape parameter of Weibull law for machine m.
- σ_m: scale parameter of the Weibull law for machine m.
- $F_m(t)$: the probability of failure of machine m in the interval $[0, t]$.
- $f_m(t)$: the density function of the Weibull law as a function of time of machine m.
- tol_m: the percentage of time of postponing or preempting a PM operation. It should be determined based on the total cost of maintenance.

Routing Parameters

- $t_{i,j}$: travel time between locations i and j.
- $c_{i,j}$: cost of routing between locations i and j.
- a_i: lower bound of the time window to perform a PM operation i.
- b_i: upper bound of the time window to perform a PM operation i.
- c: fixed penalty cost per unit time for violating the upper bound of the time window b_i.

Decision Variables

- x_{ijk}: binary variable which equal 1 if the team of technicians k traverses the arc (i, j) and 0 otherwise.
- p_i: a violation variable expressing the amount of violation of the upper bound of the time window b_i. It is equal to $max(0, \theta_i - b_i)$.
- θ_i: the start time of the maintenance operation i.
  These variables below are variable only for the maintenance model but are constant parameters for the routing one.
- δ_m: the optimal maintenance period for machine m.
- n_m: the frequency of PM operations for machine m in the fixed horizon.
- ϕ_i: the execution date of operation i.

3.2   The Maintenance Model

The objective of the maintenance model is to determine for each machine m the optimal time period δ_m to perform a PM operation while minimizing the total maintenance cost. Consequently, we obtain for each machine m the frequency n_m of PM operations. In the following, we introduce some notation used in the maintenance model.

The probability of failure of machine m in the interval $[0, \delta_m]$, where T_m is a random variable of time to failure of machine m, P_m a probability and f_m the density function of machine m, is:

$$F_m(\delta_m) = P_m(T_m \leq \delta_m) = \int_0^{\delta_m} f_m(t)\,\mathrm{d}t \qquad (1)$$

The maintenance model widely used in the literature to determine the best times to perform preventive maintenance operations for time directed tasks, is presented in [10]. Given a machine m, we look for the best time δ_m to perform preventive maintenance while minimizing the total maintenance cost $CM_m(\delta_m)$:

$$CM_m(\delta_m) = \frac{E[CM_m(\delta_m)]}{E[T_m(\delta_m)]} = \frac{Cpm_m(1 - F_m(\delta_m)) + Ccm_m F_m(\delta_m)}{\delta_m(1 - F_m(\delta_m)) + M_m(\delta_m)F_m(\delta_m)} \qquad (2)$$

The term $E[CM_m(\delta_m)]$ represents the total expected cost of a cycle of a machine m and $E[T_m(\delta_m)]$ is the expected cycle length of a machine m.

By adding the expression of the waiting cost as well as the preventive and corrective maintenance durations which are not negligible in large scales systems, Lopez-Santana et al. [6] consider this extended maintenance cost:

$$CM_m(\delta_m) = \frac{Cpm_m(1 - F_m(\delta_m)) + (Ccm_m + W_m * Cw_m)F_m(\delta_m)}{(\delta_m + Tpm_m)(1 - F_m(\delta_m)) + (M_m(\delta_m) + W_m + Tcm_m)F_m(\delta_m)} \qquad (3)$$

Where $M_m(\delta_m)$ is the expected time to failure for machine m assuming the failure occurs before δ_m.

$$M_m(\delta_m) = \int_0^{\delta_m} \frac{t f_m(t)}{F_m(\delta_m)}\,\mathrm{d}t \qquad (4)$$

Lopez-Santana et al. [6] define the waiting time W_i for an operation i as the difference between the start time of the PM operation θ_i and the expected time of failure assuming the failure occurs before θ_i:

$$W_i = \theta_i - M_m(\theta_i) \quad i \in \{1, \ldots, n_m\} \qquad (5)$$

Note that the value of θ_i is obtained from the routing model. The waiting time W_m associated to machine m is then updated considering the average value of all W_i such as i corresponds to the indexes of PM operations associated to machine m. Since the value of θ_i are unknown before solving the routing model, we consider that a PM operation will necessarily be planned at each δ_m period. The waiting time can be approximated as follows:

$$W_m = \delta_m - M_m(\delta_m) \tag{6}$$

The maintenance cost for a machine m is finally equal to:

$$CM_m(\delta_m) = \frac{Cpm_m(1 - F_m(\delta_m)) + (Ccm_m + (\delta_m - M_m(\delta_m))Cw_m)F_m(\delta_m)}{(\delta_m + Tpm_m)(1 - F_m(\delta_m)) + (\delta_m + Tcm_m)F_m(\delta_m)} \tag{7}$$

The optimal time period for each machine m corresponds to $\delta_m^* = \text{argmin } CM_m(\delta_m)$. The frequency of a maintenance operation on the planning horizon H is calculated as follows:

$$n_m = \frac{H}{E[T_m(\delta_m^*)]} \tag{8}$$

The frequency of an operation represents the number of times it must be executed in the horizon. If we have a frequency n_m, the PM operations must be executed on machine m at times $\{\delta_m^*, \delta_m^* + E[T_m(\delta_m^*)], \delta_m^* + 2 * E[T_m(\delta_m^*)], \ldots, \delta_m^* + (n_m - 1) * E[T_m(\delta_m^*)]\}$. The execution date ϕ_i of an operation i corresponding to machine m, is calculated as follows:

$$\phi_i = \delta_m^* + (i - 1) * E[T_m(\delta_m^*)], \quad i \in \{1, \ldots, n_m\} \tag{9}$$

The time windows $[a_i, b_i]$ of PM operations associated to machine m depend on $[a_m, b_m]$ the time windows for operations of machine m on the first cycle. $[a_m, b_m]$ are based on the percentage of time of postponing or preempting PM operations that we can accept. In this interval, the cost remains relatively low and near to the optimal minimal value $CM_m(\delta_m^*)$. $[a_i, b_i]$ can be determined as follows:

$$a_i = a_m + (i - 1) * E[T_m(\delta_m^*)], \quad i \in \{1, \ldots, n_m\} \tag{10}$$

$$b_i = b_m + (i - 1) * E[T_m(\delta_m^*)], \quad i \in \{1, \ldots, n_m\} \tag{11}$$

Where:

$$a_m = \delta_m^* - tol_m * \delta_m^*, \quad m \in \mathcal{M} \tag{12}$$

$$b_m = \delta_m^* + tol_m * \delta_m^*, \quad m \in \mathcal{M} \tag{13}$$

3.3   The Routing Model with Maintenance Considerations

The mathematical model can be formulated as follows:

$$\min f(x_{ijk}, \theta_i, p_i) \tag{14}$$

S.t.

$$\sum_{j=1}^{n+1} \sum_{k=1}^{K} x_{ijk} = 1, \quad \forall i \in \mathcal{O}, i \neq j \tag{15}$$

$$\sum_{i=0}^{n}\sum_{k=1}^{K}x_{ijk} = 1, \quad \forall j \in \mathcal{O}, i \neq j \tag{16}$$

$$\sum_{i=0}^{n}x_{ijk} = \sum_{i=1}^{n+1}x_{jik}, \quad \forall j \in \mathcal{O}, \forall k \in \mathcal{K} \tag{17}$$

$$\theta_i + d_i + t_{ij} \leq \theta_j + B\left(1 - \sum_{k=1}^{K}x_{ijk}\right), \quad \forall i \in \mathcal{V}^o, \forall j \in \mathcal{V}^d, i \neq j \tag{18}$$

$$a_i \leq \theta_i \leq b_i + p_i, \quad \forall i \in \mathcal{V} \tag{19}$$

$$\sum_{j=1}^{n+1}\sum_{k=1}^{K}x_{0jk} = r, \quad r \leq K \tag{20}$$

$$\sum_{i=0}^{n}\sum_{k=1}^{K}x_{in+1k} = r, \quad r \leq K \tag{21}$$

$$\sum_{j=1}^{n+1}x_{0jk} \leq 1, \quad \forall k \in \mathcal{K} \tag{22}$$

$$\theta_i, p_i \geq 0, \quad x_{ijk} \in \{0, 1\} \tag{23}$$

The objective function f depends on the variables used and equals either $f1$, $f2$ or $f3$ defined below:

$$f1 = \sum_{i=0}^{n}\sum_{j=1}^{n+1}\sum_{k=1}^{K}c_{ij}x_{ijk} + \sum_{i=1}^{n+1}c * p_i \tag{24}$$

$$f2 = \sum_{i=1}^{n}Ccm_iF_i(\theta_i) + \sum_{i=1}^{n+1}c * p_i \tag{25}$$

$$f3 = \sum_{i=1}^{n}CM_i(\theta_i) + \sum_{i=1}^{n+1}c * p_i \tag{26}$$

The objective function f1 (24) minimizes the total travel cost and the penalty's cost of violating the upper bound of the time window b_i. The objective function f2 (25) minimizes the cost of failure in order to maximize machines reliability and penalty's cost. The objective function f3 (26) minimizes the total maintenance cost and the penalty's cost. Constraints (15, 16) indicate that each operation must be performed only once. Constraints (17) ensure that if a team of technicians arrives on a vertex i to do a maintenance operation, it should leave it. The purpose of the constraints (18) is to prevent the creation of sub-tours. Constraints (19) assure that each operation is performed within its time window and set the opening time of the depot to $a_0 = 0$ and the latest possible arrival time to the depot. The violation of the closing time of the time window by the amount p_i is permitted. Constraints (20, 21) fix the number of tours by specifying that each team should go from and arrive to the depot exactly r times. Constraints (22) ensure that the teams are different whenever the tours are different and that every technicians team tour starts at the depot. Finally, the constraints (23) specify the domain values of decision variables.

In the routing model and all what follows, if the operations i correspond to the same machine's m operations then F_i, R_i, σ_i and β_i equal respectively to F_m, R_m, σ_m and β_m.

In the second objective function we have used $F_i(\theta_i)$ to exploit information from equipment degradation and therefore to consider failure risks when assigning technicians to the different tasks. This term is stochastic and non linear. It is equal to:

$$F_i(\theta_i) = 1 - e^{-(\theta_i/\sigma_i)^{\beta_i}}, \quad i \in \mathcal{O} \tag{27}$$

The density function of the Weibull law as a function of time is given below, where γ_m is the Weibull law position parameter. It is positive if the failures can not occur before the age γ_m. We consider it equal to 0 to start from the origin. Indeed, failures can occur at any time between 0 and γ_m.

$$f_m(t) = \frac{\beta_m}{\sigma_m}(\frac{t - \gamma_m}{\sigma_m})^{\beta_m - 1} e^{-(\frac{t - \gamma_m}{\sigma_m})^{\beta_m}} \tag{28}$$

The duration of the maintenance operations is calculated using δ_i^* the optimal preventive maintenance time for operation i:

$$d_i = Tpm_i(1 - F_i(\delta_i^*)) + Tcm_i F_i(\delta_i^*), \quad i \in \mathcal{O} \tag{29}$$

4   Variable Neighborhood Search

Variable neighborhood search (VNS) is a metaheuristic proposed by Mladenovic and Hansen [9]. VNS systematically changes neighborhood structures during the search for an optimal (or near-optimal) solution. Starting from an incumbent solution S, basic VNS iteratively applies two important mechanisms: a perturbation (or shaking), procedure followed by local search to improve the current solution. Shaking is essential for VNS schema as it is used to escape from the local minima traps (diversification). The local search procedure (intensification phase) aims to exploit the accumulated search experience. In this paper, we apply the most used variant of VNS, the General Variable Neighborhood Search (GVNS) which uses Variable Neighborhood Descent (VND) as a local search to explore several neighborhood structures. We have also used the Best Improvement Local Search (BILS) with the first neighborhood generated using the swap operator and the VND for our problem.

Algorithm 1. General Variable Neighborhood Search

---

1 **Initialization**
2 $\{N_1, N_2, \ldots, N_{k_{max}}\}$: Set of neighborhood structures to be used in VNS;
3 $\{N_1, N_2, \ldots, N_{l_{max}}\}$: Set of neighborhood structures to be used in VND;
4 nS : diversification parameter representing the number of times shaking is applied to the solution;
5 $S \leftarrow$ an initial solution (random or constructed for the first objective and using the constructive greedy heuristics for the others).
6 **repeat**
7     $k \leftarrow 1$;
8   **while** $k \leq k_{max}$ **do**
9        $S' \leftarrow S$;
10   **for** $p = 1 \quad to \quad nS$ **do**
11           $S' \leftarrow$ **Shaking**(S', k);
12        $S'' \leftarrow$ **VND**(S', l_{max});
13   **if** $f(S'') < f(S)$ **then**
14           $S \leftarrow S''$;
15           $k \leftarrow 1$;
16   **else**
17           $k \leftarrow k + 1$;
18 **until** *stopping criterion (maximum number of iterations)*;

4.1   Representation of the Solution and Constraints Handling

In our implementation, a solution is represented as a sequence of PM operations in each tour. We allow that a team technician arrives to an operation i after the closing of its time window. All the other constraints, including the respect of the lower bound of the time window, are hard. We therefore accept only feasible solutions that respect these hard constraints.

4.2   Greedy Constructive Heuristics for the Initial Solutions

We propose a fast heuristic to construct a good quality solution in the case where the objective function of the model is (25). This objective aims to minimize the probability that a machine fails before the start time of the preventive maintenance operation. We call therefore the heuristic, the Greedy Constructive Heuristic Failure (GCHF). The heuristic is based on the fact that the maintenance operations must be carried out as soon as possible in case of minimization of equipment's failures. The best start time an operation tends towards is the lower bound of its time window. We have also proposed a second fast heuristic to construct a good quality solution in the case where the objective function of the model is (26) which consists on the minimization of the total maintenance cost that we call the Greedy Constructive Heuristic Maintenance (GCHM). It is based on the principle that the best time of execution ϕ_i of maintenance operations is the optimal value obtained with the maintenance model. Since the routing model has the same objective, with some flexibility allowed by the time windows and more constraints due to the large number of maintenance operations, it can be a good heuristic for the problem. When using these heuristics to build an initial solution, computational time is considerably reduced for the VND and BILS. These greedy heuristics can also lead to optimal solutions on the small instances and when the penalty's cost equal 0. GCHM produces better results than GCHF for the third objective since it is dedicated to it. The heuristics are presented on the algorithm below.

Algorithm 2. Greedy Constructive Heuristics Failure and Maintenance

---

1 **Initialization** Initialize r empty routes, the lower bounds of the time window a_i and the optimal execution times ϕ_i $i \in \mathcal{O}$;
2 **Sort** a_i for GCHF (resp. ϕ_i for GCHM) from least to highest;
3 $OS \leftarrow$ Operations sorted by a_i for GCHF (resp. ϕ_i for GCHM);
4 **if** $r = 1$ **then**
5 | Insert first and last depot and this is the final solution ;
6 **else**
7 | **repeat**
8 | | Select r operations with the least a_i for GCHF (resp. ϕ_i for GCHM) from OS not previously assigned;
9 | | Insert them in the following empty positions of the r routes from first route to last route (1 to r);
10 | **until** $OS = \varnothing$;
11 | Insert first and last depot in all routes and this is the final solution;

4.3 Variable Neighborhood Descent, Best Improvement Local Search and Shaking

The Variable Neighborhood Descent is a search procedure that uses multiples neighborhoods unlike a local search which explores only one neighborhood at time.

Algorithm 3. Variable Neighborhood Descent

---

1 **Initialization**
2 $\{N_1, N_2, \ldots, N_{l_{max}}\}$: Set of neighborhood structures;
3 $S \leftarrow$ initial solution
4 **repeat**
5 | $l \leftarrow 1$;
6 | **while** $l \leq l_{max}$ **do**
7 | | Choose S' in a neighborhood N_l, such that $f(S') \leftarrow \underset{S \in N_l(S)}{\mathrm{argmin}} f(S)$;
8 | | **if** $f(S') < f(S)$ **then**
9 | | | $S \leftarrow S'$;
10 | | | $l \leftarrow 1$;
11 | | **else**
12 | | | $l \leftarrow l + 1$;
13 **until** *no improvement is obtained*;

The shaking procedure consists of selecting a random solution S' from the current k^{th} neighborhood of the current solution S ($S' \in N_k(S)$) to diversify the search process. It is applied nS times.

The best improvement local search is used to classify the neighborhoods, in the VND and as a heuristic approach.

Algorithm 4. Local Search Best Improvement

---

1 **Initialization** Choose an initial solution S ;
2 Let N be a neighborhood of S ;
3 **repeat**
4 | Choose S' in a neighborhood N of S, such that $f(S') \leftarrow \underset{S \in V(S)}{\mathrm{argmin}} f(S)$;
5 | **if** $f(S') < f(S)$ **then**
6 | | $S \leftarrow S'$;
7 **until** *S is a local optimum*;

4.4   Neighborhood Structures

The VNS and the VND procedures depend on the set of neighborhood structures chosen and the order of their execution. The neighborhood structures should be applied from the best to the least performing [9]. The sequence of moves operators that we considered is the swap, the insert, the 2-opt* and the 2-opt procedure. The same order is usually used for VRP problems in the literature.

To validate this order, we used a Steepest Descent Heuristic, known also as Best Improvement Local Search. Moves can be performed on a single route (intra-route moves) or on multiple routes (inter-route moves). For each operator used, all possible positions are examined.

The Swap move consists of exchanging two operations from the same route or from different routes.

The Insert move generates a neighbor of a solution by removing an operation from its position and inserting it in a different one, within the same route or in another route.

The 2-opt* operator removes arcs $(i, i+1)$ and $(j, j+1)$ from two different routes and reconnects arcs $(i, j+1)$ and $(j, i+1)$. This operator is inter-route.

The 2-opt operator removes arcs $(i, i+1)$ and $(j, j+1)$ from the same route and reconnects arcs (i, j) and $(i+1, j+1)$. This operator is equivalent to reversing the elements between i and $j+1$. This operator is intra-route.

The insert and 2-opt* operators change the number of operations in the routes in addition to the order of operations in the routes.

5   Computational Results

In this section, we present results of the mathematical model with the different objectives, using CPLEX solver, BILS, VND and GVNS. We also present the solutions obtained by the Greedy Constructive Heuristics proposed (GCHF and GCHM). All tests in this work were run using Python 3.6 to solve the maintenance model and using C++ for the routing one, on a windows 7, 64-bit machine, with an intel i7-4510U processor (2*2.60 GHz) and 8 GB of RAM. The solver used is CPLEX 12.7.1.

5.1   Instances Description

We use first, instances created whose values are nearer to industrial reality denoted by Re. We incorporated larger and shorter maintenance duration (Tcm and Tpm can vary from 0.5 to 48 h). We consider β superior to 1 to consider old machines where the risk of failure increase rapidly over time (wearout). We have chosen very short distances between operations (27 km) as well as very large ones (up to 500 km). The speed is 60 km/h which corresponds to reality. The horizon is fixed to $H = 101$ h for n = 12 and H = 80 h for n = 6. The penalty cost is set to $c = 10$. The depot's time window is $a_0 = 0$ and $b_{n+1} = H$. The data used to illustrate our results are presented in Table 1. In this class of instances, we consider 6 machines which can need more than one operation in the horizon. The travel time between the same machine's operations in this case is 0. We then used the first 10 machines of Solomon's instance C101 as a time matrix with a unitary speed and an horizon H of 100 h. The opening time is $a_0 = 0$ and the latest possible arrival time to the depot is $b_{n+1} = 200$. The maintenance parameters were generated in the same way as described in [6]: $Tpm_i \sim U_c[5, 10]$, $Tcm_i \sim U_c[15, 30]$, $Cpm_i \sim U_c[100, 200]$, $Ccm_i \sim U_c[400, 800]$, $Cw_i \sim U_c[10, 20]$,

$\sigma \sim t_{0i} * U_c[2,5]$, $\beta = 2$. In both class of instances, the percentage of tolerance for time windows is fixed to $tol = 7\%$. It is inspired from the values adopted in heavy industries.

Table 1. Travel time (hours) between operations and data of the illustrative case study Re

t_{ij}	0	1	2	3	4	5	6	Cpm_i	Ccm_i	Cw_i	Tpm_i	Tcm_i	$f_i(t)$
0	0	0.45	0.83	0.67	1		0.67	7.5	–	–	–	0	0	–
1	0.45	0	1	1.33	0.67	1.17	7.5	193	425	19	1	2	$W(66,2)$
2	0.83	1	0	0.33	1.67	1.17	6	156	561	14	1	2	$W(100,2)$
3	0.67	1.33	0.33	0	0.83	1.17	5.5	138	561	13	1	3	$W(63,2)$
4	1	0.67	1.67	0.83	0	0.83	5	163	462	15	0.5	3	$W(84,2)$
5	0.67	1.17	1.17	1.17	0.83	0	4	200	400	20	0.5	3	$W(90,2)$
6	7.5	7.5	6	5.5	5	4	0	600	1000	30	24	48	$W(110,2)$

5.2 Numerical Results

We have set as stopping condition for the GVNS a maximum number of iterations $Iter_{max}$. For the two first objective (24) and (25), $Iter_{max}$ equals 8 for 23 operations and 5 routes and 1 for all the other instances. For the last objective (26), $Iter_{max}$ equals 8 for 23 operations and 5 routes and 2 for all the other instances. The diversification parameter nS in the shaking phase is set to 3. In Table 3, VND, GVNS*, and GVNS represent the results of VND, the best value of GVNS and the average value of GVNS in 5 runs, respectively. We indicate also a Value Best which is the CPLEX value for the linear objective function (24) and the Best value of GVNS* for all the initial solutions when we fix the $Iter_{max}$ to 500 iterations for the non linear two others objectives. An asterisk is used to mark the optimal solutions obtained by CPLEX. In Table 2, LS(F), LS(M) and LS(R) refer respectively to BILS applied with the GCHF, GCHM and the random initial solutions.

We tested several instances to compare the different objective functions. The results are reported in Tables 2 and 3. The greedy heuristics find very near optimal solution for instances with small values of penalties in the case of minimizing the failure risk with the cost objective (25) and for the maintenance cost objective (26). They are faster than the VND and GVNS, even when using them as initial solutions with VND and GVNS. The BILS provides a good solution when choosing the best neighborhood in the search procedure and the constructed initial solutions, and is faster than VND and GVNS. The algorithms for the two first objectives are faster than the algorithms considering the third objective.

The time search of VND, BILS are inferior to the GVNS time for the model with the three objectives. We notice that for the two last objectives (25) and (26) for which initial solutions were proposed, VND is sufficient for finding the optimal solutions. For BILS, VND and GVNS, the quality of the final solutions increases with the proposed initial solutions in comparison to using random initial solutions. The execution times with the constructed initial solutions are also less than the execution times when starting from random solutions as one can notice in Table 3. Choosing a good initial solution is very important for reducing the execution time and improving rapidly the quality of the final solution. When using the greedy initial solutions, the execution time of our search algorithms considerably decreases.

The percent decrease of CPU time when using the constructive initial solution CS compared to a same random feasible initial solution RS is given by:

$$TimeDev(RS, CS) = (\frac{Time(RS) - Time(CS)}{Time(RS)}) * 100\% \qquad (30)$$

The gap between the objective value of the heuristic and the optimal solution or best solution is calculated as follows:

$$ObjDev(Best, Heuristic) = (\frac{Value(Heuristic) - Value(Best)}{Value(Best)}) * 100\% \quad (31)$$

Table 2. Results of the BILS with the first neighborhood

Inst	n	r			Routing OF				Failure OF				Maintenance OF		
				Best	LS(F)	LS(M)	LS(R)	Best	LS(F)	LS(M)	LS(R)	Best	LS(F)	LS(M)	LS(R)
Re	6	2	Value	110.88*	**110.88**	**110.88**	110.88	869.471	**869.471**	**869.471**	**869.471**	45.0156	**45.0156**	**45.0156**	45.0352
			Time	-	0.02	0.017	0.032	-	0.017	0.018	0.021	-	0.021	0.02	0.024
			ObjDev	-	0	0	0	-	0	0	0	-	0	0	0.04
			TimeDev	-	37.5	46.88	-	-	19.05	14.29	-	-	12.50	16.67	-
Re	12	4	Value	353.52*	382.8	378.96	397.68	2841.77	**2841.77**	**2841.77**	**2841.77**	100	100.02	100.02	100.064
			Time	-	0.027	0.026	0.039	-	0.025	0.026	0.035	-	0.043	0.045	0.057
			ObjDev	-	8.28	7.20	12.49	-	0	0	0	-	0.02	0.02	0.06
			TimeDev	-	30.77	33.33	-	-	28.57	25.71	-	-	24.56	21.05	-
C101	23	5	Value	225.078*	228.811	231.835	304.42	8097.44	**8097.44**	8100.5	8272.59	350.039	350.076	351.501	426.979
			Time	-	0.065	0.081	0.133	-	0.067	0.079	0.156	-	0.19	0.15	0.57
			ObjDev	-	1.226	2.564	34.67	-	0	0.037	2.163	-	0.010	0.417	21.98
			TimeDev	-	51.13	39.10	-	-	57.05	49.36	-	-	66.67	73.68	-
C101	23	8	Value	128.359*	132.027	131.671	137.131	7707.42	**7707.42**	**7707.42**	**7707.42**	235.46	**235.46**	235.472	235.464
			Time	-	0.098	0.095	0.129	-	0.051	0.056	0.134	-	0.15	0.153	0.34
			ObjDev	-	2.857	2.58	6.833	-	0	0	0	-	0	0.005	0.001
			TimeDev	-	24.03	26.36	-	-	61.94	58.21	-	-	55.88	55	-

Table 3. Results of the GVNS and the VND

Inst	n	r		Best	CIS with GCHF			CIS with GCHM			Random IS			GCH	
					VND	GVNS*	GVNS	VND	GVNS*	GVNS	VND	GVNS*	GVNS	GCHF	GCHM
								The routing cost objective function							
Re	6	2	Value	110.88*	110.88	110.88	110.88	110.88	110.88	110.88	110.88	110.88	110.88	130.8	130.8
			Time	-	0.018	0.02	0.0408	0.019	0.019	0.0378	0.021	0.02	0.041	0.017	0.016
			ObjDev	-	0	0	0	0	0	0	0	0	0	17.97	17.97
			TimeDev	-	14.29	0	0.49	9.52	5	7.80	-	-	-	-	-
Re	12	4	Value	353.52*	353.52	353.52	353.52	378.96	353.52	353.52	377.76	353.52	353.52	622.8	590.88
			Time	-	0.03	0.052	0.0558	0.026	0.055	0.0738	0.034	0.055	0.0786	0.025	0.025
			ObjDev	-	0	0	0	7.20	0	0	6.86	0	0	76.17	67.14
			TimeDev	-	11.76	5.45	29.01	23.53	0	6.11	-	-	-	-	-
C101	23	5	Value	225.078*	225.078	225.078	225.078	225.078	225.078	225.078	228.432	225.078	226.517	661.591	630.193
			Time	-	0.089	1.31041	1.72849	0.106	1.57561	1.63801	0.172	1.71601	1.89385	0.039	0.043
			ObjDev	-	0	0	0	0	0	0	1.059	0	0.211	192.69	178.79
			TimeDev	-	48.26	23.64	8.73	38.37	8.18	13.51	-	-	-	-	-
C101	23	8	Value	128.359*	129.251	128.359	128.359	128.359	128.359	128.359	128.359	128.359	128.359	149.815	150.179
			Time	-	0.085	0.249602	0.443043	0.098	0.327602	0.361922	0.168	0.405603	0.508563	0.038	0.046
			ObjDev	-	0.694	0	0	0	0	0	0	0	0	16.715	16.999
			TimeDev	-	49.4	38.46	12.88	41.67	19.23	28.83	-	-	-	-	-
								The failure cost objective function							
Re	6	2	Value	869.471	869.471	869.471	869.471	869.471	869.471	869.471	869.471	869.471	869.471	869.471	869.471
			Time	-	0.018	0.019	0.0446	0.023	0.02	0.038	0.036	0.024	0.0484	0.017	0.018
			ObjDev	0	0	0	0	0	0	0	0	0	0	0	0
			TimeDev	-	50	20.83	7.85	36.11	16.67	21.49	-	-	-	-	-
Re	12	4	Value	2841.77	2841.77	2841.77	2841.77	2841.77	2841.77	2841.77	2841.77	2841.77	2841.77	2851.67	2851.67
			Time	-	0.034	0.052	0.045	0.056	0.045	0.0608	0.08	0.055	0.0724	0.024	0.026
			ObjDev	0	0	0	0	0	0	0	0	0	0	0.35	0.35
			TimeDev	-	57.5	5.45	37.85	29.50	18.18	16.02	-	-	-	-	-
C101	23	5	Value	8097.44	8097.44	8097.44	8097.44	8100.5	8097.44	8097.44	8272.59	8097.44	8106.71	8755.39	8664.85
			Time	-	0.093	2.82362	3.31034	0.097	2.58962	2.86106	0.195	2.83922	3.48506	0.044	0.043
			ObjDev	-	0	0	0	0.037	0	0	2.163	0	0.114	8.125	7.007
			TimeDev	-	52.31	0.55	5.01	50.26	8.79	17.91	-	-	-	-	-
C101	23	8	Value	7707.42	7707.42	7707.42	7707.42	7707.42	7707.42	7707.42	7707.42	7707.42	7707.42	7707.42	7707.42
			Time	-	0.071	0.202801	0.252722	0.064	0.187201	0.287042	0.191	0.249601	0.386882	0.037	0.042
			ObjDev	-	0	0	0	0	0	0	0	0	0	0	0
			TimeDev	-	62.83	18.75	34.68	66.49	25	25.81	-	-	-	-	-
								The maintenance cost objective function							
Re	6	2	Value	45.0156	45.0156	45.0156	45.0156	45.0156	45.0156	45.0156	45.0352	45.0156	45.0156	45.0521	45.0521
			Time	-	0.021	0.035	0.0382	0.021	0.031	0.048	0.096	0.043	0.0794	0.02	0.018
			ObjDev	-	0	0	0	0	0	0	0.04	0	0	0.08	0.08
			TimeDev	-	78.13	18.60	51.89	78.13	27.91	39.55	-	-	-	-	-
Re	12	4	Value	100	100.02	100	100	100.02	100	100	100	100	100	100.132	100.132
			Time	-	0.056	0.227	0.2404	0.059	0.237	0.2702	0.068	0.279	0.2858	0.023	0.025
			ObjDev	-	0.02	0	0	0.02	0	0	0	0	0	0.13	0.13
			TimeDev	-	17.65	18.64	15.89	13.24	15.05	5.46	-	-	-	-	-
C101	23	5	Value	350.039	350.076	350.039	350.054	350.076	350.039	350.054	350.039	350.039	350.054	780.088	747.736
			Time	-	0.28	5.05443	5.68156	0.31	8.81406	10.0309	0.86	9.36006	11.9497	0.04	0.04
			ObjDev	-	0.01	0	0.042	0.01	0	0.042	0	0	0.042	122.857	113.615
			TimeDev	-	67.44	46	52.45	63.95	5.83	16.06	-	-	-	-	-
C101	23	8	Value	235.46	235.46	235.46	235.46	235.46	235.46	235.46	235.46	235.46	235.46	235.505	235.505
			Time	-	0.2	1.23241	1.49137	0.25	1.66921	1.93753	0.45	1.79401	2.12161	0.04	0.04
			ObjDev	-	0	0	0	0	0	0	0	0	0	0.019	0.019
			TimeDev	-	55.56	31.30	29.71	44.44	6.96	8.68	-	-	-	-	-

6   Conclusion

In this paper, a new formulation of the joint maintenance and routing problem is proposed. A mathematical model with different objectives functions is presented and discussed. The model consider several routes and flexible time windows which are interesting features of real industrial problems. We differentiate between essential conflicting objectives: the routing cost, the failure cost, and the total maintenance cost. The contributions of this paper are threefold: we first have introduced a non linear and stochastic failure cost that uses information from equipment degradation, with the routing model. This objective is particularly valuable for industries, where failures have serious damages. Five heuristic approaches were then proposed and compared: BILS, VND, GVNS and two proposed greedy constructive heuristics. We finally show that the quality of the initial solution, obtained with the greedy constructive heuristics, considerably reduces the CPU time and improves the objective value when using BILS, VND and GVNS. Future research will include testing these algorithms on larger instances, extending the model to a dynamic setting where incidents appear over time, and adding interesting features like skills and level of expertise.

References

1. Cordeau, J.-F., Laporte, G., Pasin, F., Ropke, S.: Scheduling technicians and tasks in a telecommunications company. J. Sched. **13**(4), 393–409 (2010)
2. Kovacs, A.A., Parragh, S.N., Doerner, K.F., Hartl, R.F.: Adaptive large neighborhood search for service technician routing and scheduling problems. J. Sched. **15**(5), 579–600 (2011)
3. Zamorano, E., Stolletz, R.: Branch-and-price approaches for the Multiperiod Technician Routing and Scheduling Problem. Eur. J. Oper. Res. **257**(1), 55–68 (2017)
4. Mathlouthi, I., Gendreau, M., Potvin, J.-Y.: Mixed integer linear programming for a multi-attribute technician routing and scheduling problem. INFOR **56**(1), 33–49 (2017)
5. Pillac, V., Guéret, C., Medaglia, A.L.: On the dynamic technician routing and scheduling problem. Research Report, École des Mines de Nantes (2012)
6. Lopez-Santana, E., Akhavan-Tabatabaei, R., Dieulle, L., Labadie, N., Medaglia, A.L.: On the combined maintenance and routing optimization problem. Reliab. Eng. Syst. Saf. **145**, 199–214 (2016)
7. Jbili, S., Chelbi, A., Radhoui, M., Kessentini, M.: Integrated strategy of Vehicle Routing and Maintenance. Reliab. Eng. Syst. Saf. **170**, 202–214 (2018)
8. Chen, Y., Polack, F., Cowling, P., Mourdjis, P., Remde, S.: Risk driven analysis of maintenance for a large-scale drainage system. In: Proceedings of 5th ICORES, pp. 296–303. SCITEPRESS - Science and Technology Publications, UK (2016)
9. Hansen, P., Mladenovic, N.: A Tutorial on Variable Neighborhood Search. G-2003-46. Les Cahiers du GERAD, Canada (2003)
10. Tsang, A.H.C.: Condition-based maintenance: tools and decision making. J. Qual. Mainten. Eng. **1**(3), 3–17 (1995)

Variable Neighborhood Search for Identical Parallel Machine Scheduling Problem with a Single Server

Abdelhak Elidrissi[1,3](✉) [ID], Mohammed Benbrahim[1], Rachid Benmansour[2,3] [ID], and David Duvivier[3] [ID]

[1] MOAD6 Team, MASI Laboratory, Mohammadia School of Engineers,
Mohammed V University, Rabat, Morocco
[2] SI2M Laboratory, National Institute of Statistics and Applied Economics,
Rabat, Morocco
r.benmansour@insea.ac.ma
[3] LAMIH UMR CNRS 8201, Polytechnic University of Hauts-de-France,
Valenciennes, France
{abdelhak.elidrissi,david.duvivier}@uphf.fr

Abstract. In this paper, we study the m identical parallel machines scheduling problem with a single server to minimize the schedule length (makespan). Each job requires a prior set-up which must be performed by a single server. For this strongly $\mathcal{NP}$-hard problem, a variable neighborhood search is proposed. We conduct a comparative analysis with existing algorithms using previously solved instances from the literature. The results indicate that the algorithm presented in this paper is effective and efficient regarding the quality of the solution: the obtained objective function values are very close to lower bounds.

Keywords: Identical parallel machines scheduling · Scheduling with a single server · Variable neighborhood search

1 Introduction

In this paper, we study the identical parallel machines scheduling problem with a single server in order to minimize the makespan. In this problem, a set $N = \{1, 2, \ldots, n\}$ of n independent jobs has to be processed on m identical parallel machines. Each jobs i is available at the beginning of the scheduling period, and has a known integer processing time p_i. Before its processing, job j must be set up on a particular machine by a single server. The setup operation, has a known integer value s_j. During the setup operation, both the machine and the server are occupied and after setting up a job on a particular machine the server becomes available. The processing operation starts immediately after the end of setup. The machines are ready at the beginning of the scheduling period. In the literature, this problem is referred to as $Pm, S1|p_j, s_j|C_{max}$, where m represents

© Springer Nature Switzerland AG 2020
R. Benmansour et al. (Eds.): ICVNS 2019, LNCS 12010, pp. 112–125, 2020.
https://doi.org/10.1007/978-3-030-44932-2_8

the number of identical parallel machines, $S1$ represents the only available server, p_j and s_j are respectively the processing time of job j and its setup time and C_{max} is the makespan ([21]). The $Pm, S1|p_j, s_j|C_{max}$ was proven to be unary $\mathcal{NP}$-hard (see [10,21]).

A feasible schedule of the $Pm, S1|p_j, s_j|C_{max}$ is a permutation π of the set N where $\pi = (\pi_1, \pi_2, \ldots, \pi_n)$, such that the jobs are scheduled according to both the availability of machines and the server. Due to the complexity of the considered problem, only small-size instances can be solved optimally by exact methods such as mixed integer programming formulation (see [13,18]). This is why we propose a metaheuristic to solve large-size instances of the $Pm, S1|p_j, s_j|C_{max}$.

We provide hereafter a brief overview of previous related works. In 2002, Abdekhodaee and Wirth [1] proposed a mixed integer programming model for the regular case and two forward/backward heuristics for the general case of the $P2, S1|p_j, s_j|C_{max}$. Abdekhodaee et al. [2] showed that the $P2, S1|p_j = p, s_j = s|C_{max}$ is $\mathcal{NP}$-hard and proposed two heuristics for the cases with equal setup time and equal processing time. Abdekhodaee et al. [3], developed a genetic algorithm, two greedy heuristics, and a version of the gilmore gomory algorithm for the general case of the $P2, S1|p_j, s_j|C_{max}$.

Gan et al. [14] proposed two mixed integer programming formulations and two variants of a branch-and-price algorithm for the $P2, S1|p_j, s_j|C_{max}$.

Kim and Lee [18] presented two mixed integer programming formulations for the $Pm, S1|p_j, s_j|C_{max}$. The first one is developed to minimize the makespan, and the second one is developed to minimize the total server waiting time (i.e., the gaps between the loading of all jobs). A hybrid heuristic algorithm is also suggested to minimize the total server waiting time for small instances with up to 40 jobs.

Hasani et al. [16] considered the $P2, S1|p_j, s_j|C_{max}$. They proposed a simulated annealing and genetic algorithms to solve large-size instances of the problem with up to 1000 jobs. The results obtained are much better than all the previous heuristics and models proposed in [3,14].

In another study by Hasani et al. [17], two constructive greedy heuristics were proposed to solve very large instances with up to 10000 jobs for the same problem. Later, Arnaout [5] introduced an ant colony optimization algorithm for the $P2, S1|p_j, s_j|C_{max}$. The computational experiments showed that the proposed method outperformed the algorithms in [16], in particular for large-size instances with up to 1000 jobs.

Elidrissi et al. [13] considered the $Pm, S1|p_j, s_j|C_{max}$ with arbitrary number of machines. The authors proposed two mixed integer programming formulations to solve the problem. The results obtained are much better than the assignment and positional dates variables formulation proposed in [18]. In another study by El Idrissi et al. [12] two constructive greedy heuristics were designed for the $Pm, S1|p_j, s_j|C_{max}$ to minimize respectively the server waiting time and machine idle time. The obtained results are much better than the algorithms presented in [3] and [17].

Benmansour et al. [7] showed that the $P2, S1|p_j = p, s_j|C_{max}$ with equal processing time is equivalent to the single-machine scheduling problem with time restrictions (STR). The STR problem is a new scheduling problem studied first in [6,9].

Recently, Alharkan et al. [4] suggested a tabu search and particle swarm optimization algorithms for the $P2, S1|p_j, s_j|C_{max}$. The conducted experiments showed that the two proposed algorithms performed well especially for medium-size and large-size instances compared to the algorithms proposed in [16,17].

The rest of the paper is organized as follows: In the next Sect. 2, we present a detailed description of the proposed Variable Neighborhood Search algorithm. Section 3 describes a new lower bound for the studied problem. Computational results are discussed in Sect. 4. Finally, Sect. 5 gives some concluding remarks.

2   Variable Neighborhood Search

Variable neighborhood search (VNS) is a metaheuristic proposed by Mladenović and Hansen [23] based on a systematic change of the neighborhood structures. The changing of the neighborhoods is motivated by the following observations: (i) A local minimum found in one neighborhood structure is not necessarily a local minimum for another neighborhood structure; (ii) A global optimum is a local one for all the neighborhood structures; (iii) For many problems, the local optimums are relatively close. VNS metaheuristic and its variants or hybrids of VNS combined with other metaheuristics have been applied to $\mathcal{NP}$-hard problems in different fields such as: scheduling, supply chain, routing, maintenance, etc. [22,24–26]. Hansen et al. [15] proposed an overview of VNS applications, VNS variants and hybrids of VNS combined with other metaheuristics.

2.1   Initial Solution

Since VNS is a trajectory-based algorithm, we need to start from an initial solution. Any permutation of all jobs π defines a feasible solution for the considered $Pm, S1|p_j, s_j|C_{max}$. The jobs should be scheduled on the machine which becomes free first at the earliest time. The first m jobs are scheduled on the first m machines, and the remaining jobs of π are scheduled if both the server and a particular machine are available. The initial solution π is generated by using the Longest Processing Time (LPT) rule which was proven to be a good rule to generate initial solution for the parallel machine scheduling problem with a single server (see [12]).

2.2   Neighborhood Structures for the $Pm, S1|p_j, s_j|C_{max}$

To obtain an efficient VNS algorithm we have to decide about three things [23]: (i) The neighborhood structures to use, (ii) The order of these neighborhoods in the search process, (iii) The search strategy to use in changing neighborhoods. We proposed the following four neighborhood structures to explore the solution space for the $Pm, S1|p_j, s_j|C_{max}$.

- $\mathcal{N}_1(\pi) = Transpose(\pi)$: The neighborhood structure consists of all permutations that can be obtained by swapping two adjacent jobs in π.
- $\mathcal{N}_2(\pi) = Swap(\pi)$: The neighborhood structure consists of all solutions obtained from solution π by swapping two random jobs of π.
- $\mathcal{N}_3(\pi) = Insertion(\pi)$: The neighborhood structure consists of all solutions obtained from the solution π by inserting each job of π at the position p $(1 \leq p \leq n)$.
- $\mathcal{N}_4(\pi) = 2\text{-}opt(\pi)$: Given two random jobs π_j and π_k we reverse the order of jobs being between those two jobs.

These neighborhood structures have been used in literature to solve different single and parallel machines scheduling problems [8, 11, 19, 22]. After performing preliminary tests, the following neighborhood order was chosen in our proposed VNS: $\mathcal{N}_4(\pi)$, $\mathcal{N}_3(\pi)$, $\mathcal{N}_2(\pi)$ and $\mathcal{N}_1(\pi)$ ($k_{max} = 4$).

2.3   Shaking and Local Search

The aim of a shaking procedure used within a VNS algorithm is to hopefully escape from local minima traps. The simple shaking procedure consists of selecting a random solution from the current neighborhood of the current solution $\mathcal{N}_k(\pi)$. Algorithm 1, summarizes the steps of the shaking phase.

Algorithm 1: Shaking(π,k)

Data: Solution π and neighborhood structure $\mathcal{N}_k$
Result: Solution π
Select randomly $\pi' \in \mathcal{N}_k(\pi)$;
$\pi \leftarrow \pi'$;
return π

Algorithm 2: Local_Search(π^0,k)

Data: Solution π^0, neighborhood structure $\mathcal{N}_k$
Result: Solution π
$\pi \leftarrow \pi^0$;
Select $\pi' \in N_k(\pi)$ such that $C_{max}(\pi') = \min_{x \in N_k(\pi)} C_{max}(x)$;
if $C_{max}(\pi') < C_{max}(\pi)$ **then**
| $\pi \leftarrow \pi'$;
end
return π

The Local Search procedure receives an initial solution π^0 from the shaking procedure and tries continually to construct a new improved solution (improved neighbor) from the current solution π by exploring its neighborhood $\mathcal{N}_k(\pi)$.

This procedure returns the local optimum within the neighborhood of the solution. In our VNS, we propose to use the best improvement search strategy for each neighborhood structure. In order to compute the cost of a given sequence of the jobs π, denoted as $C_{max}(\pi)$, reader can refer to [12]. Note that $C_{max}(\pi)$ corresponds to the makespan of the sequence π. In Algorithm 2, we present the pseudocode of the local search procedure.

2.4   VNS Algorithm

The different steps of our proposed VNS algorithm for the $Pm, S1|p_j, s_j|C_{max}$ are summarized in Algorithm 3. It starts with an initial solution generated by the LPT rule. We generate a random neighbor of this solution π with respect to the correspondent neighborhood structure $\mathcal{N}_k$. Then, we apply a best improvement local search (Algorithm 2). If no improvement exists, the neighborhood structure will be changed and the VNS algorithm stops when the execution time limit is reached or if the makespan solution is equal to the proposed lower bound.

Algorithm 3: Variable Neighborhood Search

Data: An instance of $Pm, S1|p_j, s_j|C_{max}$, neighborhood structures $\mathcal{N}_k$
for $k = 1, 2, \ldots, k_{max}$, CPU_{max}: the execution time limit
Result: Solution π, $C_{max}(\pi)$
Generate an initial solution π with LPT rule;
$k \leftarrow 1$;
while $CPU < CPU_{max}$ **do**
    while $k \le k_{max}$ **do**
        $\pi' \leftarrow$ **Shaking**(π, k);
        $\pi'' \leftarrow$ **Local_Search**(π', k);
        if $C_{max}(\pi'') < C_{max}(\pi)$ **then**
            $\pi \leftarrow \pi''$;
            $k \leftarrow 1$;
        else
            $k \leftarrow k + 1$;
        end
    end
    if $C_{max}(\pi) = LB$ **then**
        $CPU \leftarrow CPU_{max}$;
    else
        $k \leftarrow 1$;
    end
end

---

3   Lower Bound

For evaluating the quality of the VNS solution, we generalize the lower bound suggested by Abdekhodaee and Wirth [1] for the $Pm, S1|p_j, s_j|C_{max}$.

Proposition 1. *Let p_i be the processing time of the job i, and s_i its corresponding setup time.*

$$LB_1 = \sum_{i \in N} s_i + \min_{i \in N} p_i \qquad (1)$$

$$LB_2 = \frac{\sum_{i \in N} s_i + p_i}{m} \qquad (2)$$

$$LB = \max(LB_1, LB_2) \qquad (3)$$

To show the effectiveness of the proposed lower bound, we have compared it with the linear relaxation of the mixed integer programming models proposed by Elidrissi et al. [13] for several instances. We have found that our lower bound is approximately equal to the linear relaxation for the majority of the cases, in particular for the time-indexed variables model.

4   Computational Results

In this section, the evaluating of the performance of the proposed VNS algorithm is conducted by computational experiments. The proposed algorithm was coded using the C++ language. We use for run a PC with 2.90 GHz Intel(R) Core(TM) i7-4600M CPU and 16 GB of RAM memory, on Windows 7 operating system.

4.1   Benchmark Description

To verify the effectiveness of the proposed algorithm (VNS) appropriately, we perform our tests on some known benchmarks. The data was generated in the same way as described firstly by Koulamas [20]. The data are generated in the uncorrelated case, where the processing time values p_j are generated from a discrete uniform distribution U(0,100) and setup time values s_j are generated from a discrete uniform distribution U(0,100L) where $L = E(s_j)/E(p_j)$ is the server load and E(x) denotes the mean of x.

4.2   Comparison Between VNS and the MIP Model

In this Subsection, we present the results obtained by the VNS algorithm and the MIP model. The MIP model used for computational results was published by Kim and Lee [18] based on assignment and positional dates variables. In order to compare the performance of the proposed VNS with the MIP model, three factors are used: a fixed number of machines $m \in \{2, 3, 4, 5\}$, a fixed number of jobs $n \in \{8, 20\}$. The server load is fixed to $L \in \{0.1, 0.5, 0.8, 1.5, 1.8, 2\}$. 5 instances are generated randomly for each combination of (n, m, L). The computational results for $L \in \{1.5, 1.8, 2.0\}$ are given in Table 1 and the computational results for $L \in \{0.1, 0.5, 0.8\}$ are given in Table 2.

The columns of each table correspond respectively to the number of jobs n, the number of machines m, the server load L, the *minimum value* (Min), the

Table 1. Comparison of the efficiency of VNS and MIP for $L \in \{1.5, 1.8, 2.0\}$

n	m	L	CPU_{MIP}			GAP_{MIP}			R_{VNS}			R_{MIP}		
			Min	Avg	Std	Min	Avg	Std	Min	Avg	Std	Min	Avg	Std
8	2	1.5	0.29	0.38	0.09	0.00	0.00	0.00	**0.00**	**1.16**	**2.59**	**0.00**	**1.16**	**2.59**
		1.8	0.21	0.33	0.12	0.00	0.00	0.00	**0.00**	**1.27**	**2.11**	**0.00**	**1.27**	**2.11**
		2.0	0.29	0.36	0.10	0.00	0.00	0.00	**0.00**	**1.13**	**2.52**	**0.00**	**1.13**	**2.52**
	3	1.5	0.25	0.29	0.05	0.00	0.00	0.00	**0.00**	**0.00**	**0.00**	**0.00**	**0.00**	**0.00**
		1.8	0.21	0.29	0.11	0.00	0.00	0.00	**0.00**	**0.5**	**0.12**	**0.00**	**0.5**	**0.12**
		2.0	0.26	0.30	0.03	0.00	0.00	0.00	**0.00**	**0.10**	**0.23**	**0.00**	**0.10**	**0.23**
20	2	1.5	1.49	768.19	1584.42	0.00	0.20	0.45	0.00	0.78	0.51	**0.00**	**0.44**	**0.45**
		1.8	0.79	1696.27	1624.99	0.00	0.45	1.00	0.00	1.51	1.75	**0.00**	**1.46**	**1.78**
		2.0	0.53	25.34	54.11	0.00	0.00	0.00	0.00	0.31	0.70	**0.00**	**0.18**	**0.41**
	3	1.5	0.29	0.49	0.15	0.00	0.00	0.00	**0.00**	**0.50**	**0.10**	**0.00**	**0.50**	**0.10**
		1.8	0.36	0.69	0.26	0.00	0.00	0.00	**0.00**	**0.10**	**0.30**	**0.00**	**0.10**	**0.30**
		2.0	0.38	0.59	0.21	0.00	0.00	0.00	**0.00**	**0.10**	**0.20**	**0.00**	**0.10**	**0.20**
	4	1.5	0.41	0.63	0.35	0.00	0.00	0.00	**0.00**	**0.80**	**0.18**	**0.00**	**0.80**	**0.18**
		1.8	0.53	0.71	0.18	0.00	0.00	**0.00**	**0.00**	**0.00**	**0.00**	**0.00**	**0.00**	**0.00**
		2.0	0.31	0.39	0.11	0.00	0.00	0.00	**0.00**	**0.00**	**0.00**	**0.00**	**0.00**	**0.00**
	5	1.5	0.28	0.51	0.15	0.00	0.00	0.00	**0.00**	**0.00**	**0.00**	**0.00**	**0.00**	**0.00**
		1.8	0.43	0.52	0.09	0.00	0.00	0.00	**0.00**	**0.00**	**0.00**	**0.00**	**0.00**	**0.00**
		2.0	0.42	0.59	0.15	0.00	0.00	0.00	**0.00**	**0.10**	**0.20**	**0.00**	**0.10**	**0.20**

average value (Avg) and the *standard-deviation value* (Std) of the CPU time
(in seconds) (over 5 test instances) needed to compute optimal values using the
MIP formulation solved by IBM ILOG CPLEX 12.6 solver on C++: (CPU_{MIP}),
the *minimum value* (Min), the *average value* (Avg) and the *standard-deviation
value* (Std) of the GAP: (GAP_{MIP}), the deviation of the VNS algorithm from
the lower bound (R_{VNS}), and in the last column, the deviation of the MIP model
from the lower bound (R_{MIP}).

The last two metrics are calculated as follows:

$$R_{VNS} = \frac{C_{max}(VNS) - LB}{LB} \times 100$$

$$R_{MIP} = \frac{C_{max}(MIP) - LB}{LB} \times 100$$

$C_{max}(VNS)$ and $C_{max}(MIP)$, are the solution values obtained by the VNS
algorithm and the MIP model.

- The time limit for CPLEX was set to 3600 s for all instances.
- The stopping criteria for VNS was set to 165 s for $L \in \{1.5, 1.8, 2.0\}$ and 650 s
  for $L \in \{0.1, 0.5, 0.8\}$.

Table 2. Comparison of the efficiency of VNS and MIP for $L \in \{0.1, 0.5, 0.8\}$

n	m	L	CPU_{MIP}			GAP_{MIP}			R_{VNS}			R_{MIP}		
			Min	Avg	Std	Min	Avg	Std	Min	Avg	Std	Min	Avg	Std
8	2	0.1	1.52	2.02	0.68	0.00	0.00	0.00	**0.00**	**0.15**	**0.20**	**0.00**	**0.15**	**0.20**
		0.5	0.80	1.00	0.15	0.00	0.00	0.00	**0.25**	**0.86**	**0.55**	0.25	0.86	0.55
		0.8	0.44	0.63	0.15	0.00	0.00	0.00	**1.81**	**4.03**	**2.25**	1.81	4.03	2.25
20	2	0.1	3600	3600	0.00	33.23	48,86	12.93	0.00	0.05	0.07	**0.00**	**0.02**	**0.04**
		0.5	3600	3600	0.00	13.52	22.89	11.29	**0.00**	**0.14**	**0.14**	0.00	0.18	0.20
		0.8	3600	3600	0.00	0.70	6.01	6.67	0.05	1.71	1.45	**0.26**	**1.54**	**1.49**

The following summary can be given:

For $L > 1$, the VNS algorithm found optimal solutions for all instances except the two cases with $(n = 20, m = 2, L = 1.5)$ and $(n = 20, m = 2, L = 1.8)$, where both CPLEX and VNS are not able to produce optimal solution for all instances. For $L < 1$, CPLEX and VNS are not able to produce optimal solution for all instances for $(n = 20, m = 2, L \in \{0.1, 0.5, 0.8\})$.

In addition, for $(n = 20, m = 2, L = 0.5)$ VNS produces better solution in comparison with the MIP model in term of the deviation from the lower bound and the average solution time of VNS for this case is 597.98 s. Due to the complexity of the $Pm, S1|p_j, s_j|C_{max}$, VNS can also be used to solve large-size instances, while the MIP model is only suitable for small-size instances. In the next subsection, we compare the results of the VNS algorithm with benchmark algorithms existing in the literature.

4.3   Comparison Between VNS and Benchmark Algorithms

In order to compare the performance of the proposed VNS with benchmark algorithms from the literature, three factors are used:

- A fixed number of machines.
- Problem size fixed to three levels: small-size instances with $n \in (8, 20)$, medium-size instances with $n \in (30, 40, 50, 100)$ and large-size instance with $n \in (200, 250, 300, 350)$.
- The server load is also fixed to $L \in \{0.1, 0.5, 0.8, 1.5, 1.8, 2\}$.

The stopping criteria for VNS was set to 165 s for $L \in \{1.5, 1.8, 2.0\}$ and to 750 s for $L \in \{0.1, 0.5, 0.8\}$.

For each value of the server load 5 instances are generated randomly for $n \in \{8, 20\}$ and 10 instances were generated randomly for each of the other combinations of n, m and L. The computational results of VNS for $L \in \{0.1, 0.5, 0.8\}$ are given in Tables 3 and 4. The computational results of VNS for $L \in \{1.5, 1.8, 2.0\}$ are given in Tables 5, 6 and 7. In Tables 3, 4, 5, 6 and 7, column 1 gives the number of jobs, column 2 gives the number of machines, columns 4 until the last column

Table 3. Computational results of VNS for small, medium and large-size instances for $L \in \{0.1, 0.5\}$

n	m		$L = 0.1$			$L = 0.5$		
			VNS	HS1-LST	FH/(MIT)	**VNS**	HS1-LST	FH/(MIT)
8	2	R_{avg}	**1.00146**	1.03591	1.03591	**1.00860**	1.17261	1.08607
		R_{max}	**1.00424**	1.11385	1.11385	**1.01645**	1.30882	1.13388
20	2	R_{avg}	**1.00047**	1.02269	1.02464	**1.00142**	1.06107	1.04708
		R_{max}	**1.00145**	1.02269	1.01634	1.00337	1.13828	**1.00260**
50	2	R_{avg}	**1.00034**	1.01033	1.01119	**1.00322**	1.02050	1.01431
		R_{max}	**1.00127**	1.03451	1.03080	**1.00482**	1.03686	1.02459
100	2	R_{avg}	**1.00033**	1.00611	1.00611	**1.00389**	1.01405	1.00910
		R_{max}	**1.00070**	1.01296	1.01296	**1.00833**	1.04478	1.03046
150	2	R_{avg}	**1.00017**	1.00459	1.00459	**1.00271**	1.00572	1.00429
		R_{max}	**1.00039**	1.00745	1.00745	**1.00409**	1.01021	1.00822
200	2	R_{avg}	**1.00010**	1.00484	1.00484	1.00371	1.00443	**1.00259**
		R_{max}	**1.00027**	1.00825	1.00825	1.00625	1.00985	**1.00518**
250	2	R_{avg}	**1.00021**	1.00251	1.00251	**1.00297**	1.00370	1.00230
		R_{max}	**1.00065**	1.00614	1.00614	**1.00424**	1.00883	1.00610
300	2	R_{avg}	**1.00007**	1.00182	1.00182	1.00279	1.00144	**1.00098**
		R_{max}	**1.00031**	1.00470	1.00470	1.00376	1.00509	**1.00262**
350	2	R_{avg}	**1.00008**	1.00090	1.00090	1.00353	1.00122	**1.00098**
		R_{max}	**1.00021**	1.00248	1.00248	1.00491	1.00395	**1.00167**

give the values R_{avg}, denoting the average value of the ratio C_{max}/LB, and R_{max}, denoting the maximum value of the ratio C_{max}/LB among all instances for a particular value of L and for a particular algorithm.

Comparing the obtained results for VNS for $L \in \{0.1, 0.5, 0.8\}$ with the results in El Idrissi et al. [12] for the HS1-LST heuristic and the results in Hasani et al. [17] for the Min-idle algorithm which we denote as (MIT) and the results in Abdekhodaee and Wirth [3] for the forward heuristic which we denote as (FH) and also comparing the obtained results of VNS for $L \in \{1.5, 1.8, 2.0\}$ with the results in El Idrissi et al. [12] for the HS2-LPT heuristic and the results in Hasani et al. [17] for the Min-loadgap algorithm which we denote as (MLG) and the results in Abdekhodaee and Wirth [3] for the backward heuristic which we denote as (BH) when using the same instances, the following summary can be given:

In Tables 3 and 4, VNS is compared with HS1-LST, FH and MIT for $L \in \{0.1, 0.5, 0.8\}$ and for $m = 2$. Since FH and MIT can be applied only for the case of two machines. It must be noted that FH and MIT provide the same result for all proposed instances. VNS algorithm outperform all algorithms for the majority of the cases. For example, for the case of ($n = 8, m = 2, L = 0.5$)

Table 4. Computational results of VNS for small, medium and large-size instances for $L = 0.8$

n	m		$L = 0.8$		
			VNS	HS1-LST	FH/(MIT)
8	2	R_{avg}	**1.04170**	1.20681	1.13497
		R_{max}	**1.06886**	1.35988	1.17917
20	2	R_{avg}	**1.01754**	1.08417	1.10650
		R_{max}	**1.03368**	1.17402	1.27185
50	2	R_{avg}	**1.01871**	1.06299	1.03064
		R_{max}	**1.03708**	1.12111	1.07610
100	2	R_{avg}	**1.02300**	1.04793	1.02527
		R_{max}	**1.03003**	1.08919	1.04048
150	2	R_{avg}	1.01478	1.01687	**1.00722**
		R_{max}	1.02487	1.03687	**1.01504**
200	2	R_{avg}	1.01652	1.01611	**1.00844**
		R_{max}	1.02379	1.02805	**1.01666**
250	2	R_{avg}	1.01532	1.01129	**1.00498**
		R_{max}	1.01942	1.02280	**1.01124**
300	2	R_{avg}	1.01894	1.02038	**1.00678**
		R_{max}	1.02265	1.03069	**1.01391**
350	2	R_{avg}	1.01895	1.01518	**1.00646**
		R_{max}	1.03576	1.03719	**1.02185**

and for VNS, the maximum value of the relation C_{max}/LB was 1.01645 and the average value for the relation C_{max}/LB was 1.005141, where for HS1-LST the maximum value of the relation C_{max}/LB was 1.30882 and the average value for the relation C_{max}/LB was 1.17261.

In Tables 6 and 7, VNS is compared with HS2-LPT, BH and MLG for $L \in \{1.5, 1.8, 2.0\}$ for small-size instances. As a point of clarification MLG was proposed only for the case of two machines and the symbol $(\star)$ is used to specify that no solution can be found with this algorithm. VNS is better than all all examined algorithms in term of the deviation from the lower bound for $(n = 8, m = 2)$, $(n = 8, m = 3)$, $(n = 20, m = 2)$. It provides also the same results as HS2-LPT for the remaining cases.

In Table 5, VNS is compared with HS2-LPT and BH for $m \geq 2$ and for $L \in \{1.5, 1.8, 2.0\}$. It can be noted that VNS algorithm is able to produce optimal solution for all combinations of $(n = 30, m = 3)$, $(n = 40, m = 4)$, $(n = 40, m = 5)$, $(n = 50, m = 3)$, $(n = 50, m = 4)$, $(n = 100, m = 3)$, $(n = 100, m = 5)$, $(n = 200, m = 3)$, $(n = 200, m = 4)$, $(n = 250, m = 3)$ and $(n = 2500, m = 5)$.

The overall results show that the performance of the VNS algorithm is related to the number of jobs, the number of machines and the the value of the server

Table 5. Computational results of VNS for medium and large-size instances for $L \in \{1.5, 1.8, 2.0\}$

n	m		$L = 1.5$			$L = 1.8$			$L = 2.0$		
			VNS	HS2-LPT	BH	VNS	HS2-LPT	BH	VNS	HS2-LPT	BH
30	3	R_{avg}	1.00000	1.00000	1.00264	1.00000	1.00000	1.00026	1.00000	1.00000	1.00000
		R_{max}	1.00000	1.00000	1.01634	1.00000	1.00000	1.0026	1.00000	1.00000	1.00000
	4	R_{avg}	1.00014	1.00014	1.00014	1.00018	1.00018	1.00018	1.00000	1.00000	1.00000
		R_{max}	1.00141	1.00141	1.00141	1.00181	1.00181	1.00181	1.00000	1.00000	1.00000
40	3	R_{avg}	1.00000	1.00000	1.0025	1.00000	1.00000	1.00000	1.00034	1.00034	1.00070
		R_{max}	1.00000	1.00000	1.0184	1.00000	1.00000	1.00000	1.00336	1.00336	1.00336
	4	R_{avg}	1.00000	1.00000	1.00000	1.00000	1.00000	1.00000	1.00000	1.00000	1.00000
		R_{max}	1.00000	1.00000	1.00000	1.00000	1.00000	1.00000	1.00000	1.00000	1.00000
	5	R_{avg}	1.00000	1.00000	1.00000	1.00000	1.00000	1.00000	1.00000	1.00000	1.00000
		R_{max}	1.00000	1.00000	1.00000	1.00000	1.00000	1.00000	1.00000	1.00000	1.00000
50	3	R_{avg}	1.00000	1.00000	1.00145	1.00000	1.00000	1.00007	1.00000	1.00000	1.00000
		R_{max}	1.00000	1.00000	1.00848	1.00000	1.00000	1.00065	1.00000	1.00000	1.00000
	4	R_{avg}	1.00000	1.00000	1.00000	1.00000	1.00000	1.00000	1.00000	1.00000	1.00000
		R_{max}	1.00000	1.00000	1.00000	1.00000	1.00000	1.00000	1.00000	1.00000	1.00000
60	3	R_{avg}	1.00002	1.00002	1.00027	1.00000	1.00000	1.00029	1.00000	1.00000	1.00000
		R_{max}	1.00021	1.00021	1.00142	1.00000	1.00000	1.00235	1.00000	1.00000	1.00000
	5	R_{avg}	1.00002	1.00002	1.00002	1.00000	1.00000	1.00000	1.00000	1.00000	1.00000
		R_{max}	1.00021	1.00021	1.00021	1.00000	1.00000	1.00000	1.00000	1.00000	1.00000
100	3	R_{avg}	1.00000	1.00000	1.00000	1.00000	1.00000	1.00010	1.00000	1.00000	1.00000
		R_{max}	1.00000	1.00000	1.00000	1.00000	1.00000	1.00102	1.00000	1.00000	1.00000
	5	R_{avg}	1.00000	1.00000	1.00000	1.00000	1.00000	1.00000	1.00000	1.00000	1.00000
		R_{max}	1.00000	1.00000	1.00000	1.00000	1.00000	1.00000	1.00000	1.00000	1.00000
200	3	R_{avg}	1.00000	1.00000	1.00002	1.00000	1.00000	1.00000	1.00000	1.00000	1.00000
		R_{max}	1.00000	1.00000	1.00014	1.00000	1.00000	1.00000	1.00000	1.00000	1.00000
	4	R_{avg}	1.00000	1.00000	1.00000	1.00000	1.00000	1.00000	1.00000	1.00000	1.00000
		R_{max}	1.00000	1.00000	1.00000	1.00000	1.00000	1.00000	1.00000	1.00000	1.00000
250	3	R_{avg}	1.00000	1.00000	1.00006	1.00000	1.00000	1.00001	1.00000	1.00000	1.00000
		R_{max}	1.00000	1.00000	1.00051	1.00000	1.00000	1.00009	1.00000	1.00000	1.00000
	5	R_{avg}	1.00000	1.00000	1.00000	1.00000	1.00000	1.00000	1.00000	1.00000	1.00000
		R_{max}	1.00000	1.00000	1.00000	1.00000	1.00000	1.00000	1.00000	1.00000	1.00000

Table 6. Computational results of VNS for small-size instances for $L \in \{1.5, 1.8\}$

n	m		$L = 1.5$				$L = 1.8$			
			VNS	HS2-LPT	BH	MLG	VNS	HS2-LPT	BH	MLG
8	2	R_{avg}	1.01157	1.04044	1.02051	1.03388	1.01273	1.03848	1.0274	1.03959
		R_{max}	1.05785	1.16529	1.08058	1.16942	1.04874	1.11372	1.11372	1.11045
	3	R_{avg}	1.00000	1.00000	1.00000	$\star$	1.00053	1.00103	1.01303	$\star$
		R_{max}	1.00000	1.00000	1.00000	$\star$	1.00266	1.00266	1.0625	$\star$
20	2	R_{avg}	1.00777	1.02533	1.02085	1.01949	1.01506	1.04400	1.03147	1.05137
		R_{max}	1.01332	1.03900	1.03874	1.04212	1.03971	1.09224	1.07453	1.09835
	3	R_{avg}	1.00045	1.00045	1.00076	$\star$	1.00011	1.00011	1.00011	$\star$
		R_{max}	1.00226	1.00226	1.00226	$\star$	1.00057	1.00057	1.00057	$\star$
	4	R_{avg}	1.00083	1.00083	1.00547	$\star$	1.00000	1.00000	1.00000	$\star$
		R_{max}	1.00413	1.00413	1.02324	$\star$	1.00000	1.00000	1.00000	$\star$
	5	R_{avg}	1.00000	1.00000	1.00000	$\star$	1.00000	1.00000	1.00000	$\star$
		R_{max}	1.00000	1.00000	1.00000	$\star$	1.00000	1.00000	1.00000	$\star$

load. The best performance is observed by VNS algorithm. For $L \in \{1.5, 1.8, 2.0\}$, VNS is able to produce an optimal solution for the majority of the cases. For $L \in \{0.1, 0.5, 0.8\}$, VNS shows better results in term of the deviation from the lower bound in comparison with all proposed algorithms.

Table 7. Computational results of VNS for small-size instances for $L = 2.0$

n	m		$L = 2.0$			
			VNS	HS2-LPT	BH	MLG
8	2	R_{avg}	**1.01126**	1.03309	1.02827	1.03480
		R_{max}	**1.05628**	1.09957	1.09668	1.13131
	3	R_{avg}	**1.00101**	**1.00101**	1.02060	⋆
		R_{max}	**1.00504**	**1.00504**	1.05804	⋆
20	2	R_{avg}	1.00313	1.01054	**1.00295**	1.00536
		R_{max}	1.01567	1.03893	**1.01416**	1.02679
	3	R_{avg}	**1.00010**	**1.00010**	**1.00010**	⋆
		R_{max}	**1.00050**	**1.00050**	**1.00050**	⋆
	4	R_{avg}	**1.00000**	**1.00000**	**1.00000**	⋆
		R_{max}	**1.00000**	**1.00000**	**1.00000**	⋆
	5	R_{avg}	**1.00011**	**1.00011**	**1.00011**	⋆
		R_{max}	**1.00054**	**1.00054**	**1.00054**	⋆

5 Conclusion

This paper proposes a variable neighborhood search algorithm to solve small, medium and large-size instances of the arbitrary number of identical parallel machines scheduling problem with a single server to minimize the makespan. The instances were generated in the same way as in previous works so that the results can be compared with existing algorithms in the literature. Based on the generated instances, it turns out that the proposed VNS algorithm performed very well in terms of the deviation from the lower bound. In particular, it outperformed the heuristics of El Idrissi et al. [12], Abdekhodaee et al. [3] and Hasani et al. [17] and reached the lower bound for the majority of the cases for $L > 1$. In future work, it will be interesting to compare different metaheuristic approaches for solving the $Pm, S1|p_j, s_j|C_{max}$, and take into consideration other types of objective functions.

References

1. Abdekhodaee, A.H., Wirth, A.: Scheduling parallel machines with a single server: some solvable cases and heuristics. Comput. Oper. Res. **29**(3), 295–315 (2002)

2. Abdekhodaee, A.H., Wirth, A., Gan, H.S.: Equal processing and equal setup time cases of scheduling parallel machines with a single server. Comput. Oper. Res. **31**(11), 1867–1889 (2004)
3. Abdekhodaee, A.H., Wirth, A., Gan, H.S.: Scheduling two parallel machines with a single server: the general case. Comput. Oper. Res. **33**(4), 994–1009 (2006)
4. Alharkan, I., Saleh, M., Ghaleb, M.A., Kaid, H., Farhan, A., Almarfadi, A.: Tabu search and particle swarm optimization algorithms for two identical parallel machines scheduling problem with a single server. J. King Saud Univ. Eng. Sci. (2019)
5. Arnaout, J.P.: Heuristics for the two-machine scheduling problem with a single server. Int. Trans. Oper. Res. **24**(6), 1347–1355 (2017)
6. Benmansour, R., Braun, O., Artiba, A.: On the single-processor scheduling problem with time restrictions. In: 2014 International Conference on Control, Decision and Information Technologies (CoDIT), pp. 242–245. IEEE (2014)
7. Benmansour, R., Braun, O., Hanafi, S.: The single-processor scheduling problem with time restrictions: complexity and related problems. J. Sched. **22**(4), 465–471 (2018). https://doi.org/10.1007/s10951-018-0579-8
8. Benmansour, R., Braun, O., Hanafi, S., Mladenovic, N.: Using a variable neighborhood search to solve the single processor scheduling problem with time restrictions. In: Sifaleras, A., Salhi, S., Brimberg, J. (eds.) ICVNS 2018. LNCS, vol. 11328, pp. 202–215. Springer, Cham (2019). https://doi.org/10.1007/978-3-030-15843-9_16
9. Braun, O., Chung, F., Graham, R.: Single-processor scheduling with time restrictions. J. Sched. **17**(4), 399–403 (2013). https://doi.org/10.1007/s10951-013-0342-0
10. Brucker, P., Dhaenens-Flipo, C., Knust, S., Kravchenko, S.A., Werner, F.: Complexity results for parallel machine problems with a single server. J. Sched. **5**(6), 429–457 (2002)
11. De Paula, M.R., Ravetti, M.G., Mateus, G.R., Pardalos, P.M.: Solving parallel machines scheduling problems with sequence-dependent setup times using variable neighbourhood search. IMA J. Manag. Math. **18**(2), 101–115 (2007)
12. Elidrissi, A., Benbrahim, M., Benmansour, R., Duvivier, D.: Greedy heuristics for identical parallel machine scheduling problem with single server to minimize the makespan. In: MATEC Web of Conferences, vol. 200, p. 00001. EDP Sciences (2018)
13. Elidrissi, A., Benmansour, R., Benbrahim, M., Duvivier, D.: MIP formulations for identical parallel machine scheduling problem with single server. In: 2018 4th International Conference on Optimization and Applications (ICOA), pp. 1–6. IEEE (2018)
14. Gan, H.S., Wirth, A., Abdekhodaee, A.: A branch-and-price algorithm for the general case of scheduling parallel machines with a single server. Comput. Oper. Res. **39**(9), 2242–2247 (2012)
15. Hansen, P., Mladenović, N., Pérez, J.A.M.: Variable neighbourhood search: methods and applications. Ann. Oper. Res. **175**(1), 367–407 (2010)
16. Hasani, K., Kravchenko, S.A., Werner, F.: Simulated annealing and genetic algorithms for the two-machine scheduling problem with a single server. Int. J. Prod. Res. **52**(13), 3778–3792 (2014)
17. Hasani, K., Kravchenko, S.A., Werner, F.: Minimizing the makespan for the two-machine scheduling problem with a single server: two algorithms for very large instances. Eng. Optimiz. **48**(1), 173–183 (2016)
18. Kim, M.Y., Lee, Y.H.: MIP models and hybrid algorithm for minimizing the makespan of parallel machines scheduling problem with a single server. Comput. Oper. Res. **39**(11), 2457–2468 (2012)

19. Kirlik, G., Oguz, C.: A variable neighborhood search for minimizing total weighted tardiness with sequence dependent setup times on a single machine. Comput. Oper. Res. **39**(7), 1506–1520 (2012)
20. Koulamas, C.P.: Scheduling two parallel semiautomatic machines to minimize machine interference. Comput. Oper. Res. **23**(10), 945–956 (1996)
21. Kravchenko, S.A., Werner, F.: Parallel machine scheduling problems with a single server. Math. Comput. Modell. **26**(12), 1–11 (1997)
22. Krim, H., Benmansour, R., Duvivier, D., Artiba, A.: A variable neighborhood search algorithm for solving the single machine scheduling problem with periodic maintenance. RAIRO Oper. Res. **53**(1), 289–302 (2019)
23. Mladenović, N., Hansen, P.: Variable neighborhood search. Comput. Oper. Res. **24**(11), 1097–1100 (1997)
24. Perron, S., Hansen, P., Le Digabel, S., Mladenović, N.: Exact and heuristic solutions of the global supply chain problem with transfer pricing. Eur. J. Oper. Res. **202**(3), 864–879 (2010)
25. Schneider, M., Stenger, A., Hof, J.: An adaptive vns algorithm for vehicle routing problems with intermediate stops. OR Spectrum **37**(2), 353–387 (2015)
26. Todosijević, R., Benmansour, R., Hanafi, S., Mladenović, N., Artiba, A.: Nested general variable neighborhood search for the periodic maintenance problem. Eur. J. Oper. Res. **252**(2), 385–396 (2016)

Basic VNS for the Uncapacitated Single Allocation p-Hub Maximal Covering Problem

Matheus de Araujo Butinholi[(✉)], Alexandre Xavier Martins,
Paganini Barcellos de Oliveira, and Diego Perdigão Martino

Instituto de Ciências Exatas e Aplicadas, Departamento de Engenharia de Produção,
Universidade Federal de Ouro Preto, João Monlevade, Brazil
{matheus.butinholi,diego.martino}@aluno.ufop.edu.br,
{xmartins,paganini}@ufop.edu.br

Abstract. This paper addresses the Uncapacitated Single Allocation p-hub Maximal Covering Problem (USApHMCP), which aims to determine the best allocation for the p-hubs within a node network in order to maximize the network coverage. We proposed a search strategy-based heuristic Basic Variable Neighborhood Search (VNS) to solve the problem. Two different sets of test instances from the literature, Civil Aeronautics Board (CAB) and Australian Post (AP), were used to evaluate the performance of VNS and to compare it with the Tabu Search (TS) metaheuristic. In most instances, the bounds obtained by VNS and TS were the same but, on the other hand, for some of them, VNS presented a slight advantage and vice versa. That is, both algorithms are convenient to solve the proposed problem.

Keywords: Maximum coverage problem · Uncapacitated Single Allocation p-hub · Variable Neighborhood Search

1 Introduction

Hub-and-spokes networks are interconnection paths between hubs and remote nodes. Given a set of nodes, the hubs are strategically chosen based on the amount of activity/service flow intended to be transported among nodes, that is, it is necessary to establish a connection configuration network capable of covering the demands among nodes at the same time that consider the transportation costs [8]. Moreover, authors in [5] highlight that it is unfeasible to create a network that makes the direct path among all pairs of nodes due to its high cost in most of the cases.

Real applications of hub-and-spoke networks are easily found in environments such as the public transportation, telecommunication systems, logistic systems

Supported by Ufop, Fapemig, Capes and CNPq.

and air transportation [11]. In addition, authors in [5] present that hub-and-spoke problems also include satellites, industrial and postal delivery networks. In all cases, authors in [11] state that hubs are used in a way that the traffic of remote nodes is consolidated, which allows an efficient and effective connection among the arcs through simplified routes of low costs. In contrast, when the total number of hubs (p) and connections are limited, the trade-off between traffic cost and remote nodes coverage for the hubs becomes a challenge.

The study of the problem of the maximal coverage of a predefined number of p-hubs was introduced by author in [2]. In his studies, Campbell proposed four integer programming approaches and formulations for the problem: the p-median problem; the uncapacitated facility location problem; the p-center problem and the hub covering problem. The first considers opening p facilities in order to minimize the sum of the distances among the facilities and remote nodes. The second aims to minimize the sum of facilities installation and the nodes supply (customers), both predefined through installation and transportation cost parameters, considering that the facilities have an endless supply capacity. The third is a variation of the data clustering problem and aims partitioning a set of nodes in p clusters according to the distances minimization among hubs and served points. The fourth aims to maximize the p-hubs covering percentage considering their unlimited services flow capacity and only enables simple allocation among nodes.

In this sense, this paper presents the Uncapacitated Single Allocation p-Hub Maximal Covering Problem (USApHMCP), which aims to maximize the covering of the uncapacitated hubs through different arrange possibilities of connections in a network, since these connections are single allocations between each node and its respective hub. In the literature, this problem is classified as $\mathcal{NP}$-Hard, that is, the computational effort grows exponentially as the number of nodes is increased to solve the problem [6]. Thus, integer programming formulations may present difficulty to reach the optimal solutions of USApHMCP as the problem dimension grows.

Due to this difficulty of exact approach to solve the large-scale USApHMCP instances, authors in [11] have proposed an approximate approach for this problem based on the metaheuristic Tabu Search (TS), introduced by authors in [4]. TS is an adaptive procedure that guides the local searches in the neighborhoods and presents as main characteristic the use of a tabu list of prohibited solutions normally associated to the last n solutions inserted. In the TS algorithm, given an initial solution, complete descent procedures are done in a neighborhood in parallel with the addition of these solutions in the tabu list. In that way, the method prevents from visiting the recent solutions, which enables to discover new neighborhoods until satisfying a stop criteria.

In order to provide an approximated strategy not presented so far in the literature for this problem, this paper proposes the metaheuristic Basic Variable Neighborhood Search (VNS) proposed by [7] for the USApHMCP. To analyze the computational performance, two sets of instances from the literature – Civil Aeronautic Board (CAB) and Australian Post (AP) – are solved and compared with the results obtained by TS developed by [11].

Based on this methodology, this paper is divided into sections as follows: Sect. 2 describes the mathematical formulation for the USApHMCP; Sect. 3 presents the metaheuristic Basic VNS; computational experiments are in Sect. 4, while conclusions and remarks are shown in Sect. 5.

2   Uncapacitated Single Allocation p-Hub Maximal Covering Problem

Different mathematical formulations regarding the allocation of p-hubs can be found in the literature. Specifically, for this paper, authors have chosen the USApHMCP formulation proposed by [10] and presented in the next paragraphs.

Let N be a set of nodes and p the total number of hubs, W_{ij} the flow services and C_{ij} the transportation cost among the nodes, all of them for all pair $(i,j) \in N$. It is possible to state a path among the nodes $(i \to k \to m \to j)$ composed by two remote nodes i and j and two hubs k and m, since the total path cost should be lower or equal to β, that is, $\chi C_{ik} + \alpha C_{km} + \delta C_{mj} \leq \beta$ in which χ and δ are discounting factors between the hub and the spoke, and α the discount factor of paths between the hubs. All these factors are in range $[0, 1]$. To evaluate the possibility of the paths coverage, the binary parameter $a_{i,j}^{k,m}$ is inserted in the formulation.

$$a_{i,j}^{k,m} = \begin{cases} 1 : \chi C_{ik} + \alpha C_{km} + \delta C_{mj} \leq \beta \\ 0 : \text{otherwise.} \end{cases}$$

Thus, given a network of nodes, the problem consists in the efficient allocation of the spokes to the hubs, with the aim to maximize the coverage of all service flow of the network. For that, each spoke node can only be connected to only one hub. The hub, in turn, can be connected to all the remote nodes and should be connected to the all others hubs in the network. Figure 1 exemplify a possible hub-and-spoke allocation in a network nodes which contains a total of ten nodes and p equal to 2, that is, only two flow hubs are selected.

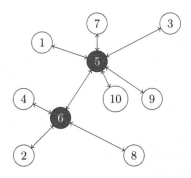

Fig. 1. Example of a hub-and-spoke network.

Figure 1 presents the simple connection of remote nodes $n = \{1, 3, 7, 9, 10\}$ to the hub flow 5 and the connection of nodes $n = \{2, 4, 8\}$ to the hub 6 at the same time. Besides that, it is necessary to emphasize that the hubs can be connected with all the spokes and have links to all the other network hubs as represented by the arc between nodes 5 and 6, aiming the transition flow connection.

In this context, adding the decision variables X_{ij} and Z_{ij} which are responsible to indicate if a node i is connected (1) to a hub j or not (0) and state the flow coverage fraction of node i to j, respectively, it is possible to structure the mathematical formulation described by Eqs. 1–7.

$$\max \sum_{i \in N} \sum_{j \in N} W_{ij} Z_{ij} \tag{1}$$

$$\text{s.a:} \sum_{k \in N} X_{kk} = p \tag{2}$$

$$\sum_{k \in N} X_{ik} = 1 \qquad\qquad \forall\, i \in N \tag{3}$$

$$X_{ik} \leq X_{kk} \qquad\qquad \forall\, i, k \in N \tag{4}$$

$$Z_{ij} \leq \sum_{k \in N} a_{i,j}^{k,m} X_{ik} + \lambda_{ij}(1 - X_{jm}) \qquad \forall\, i, j \in N, m \in N \tag{5}$$

$$X_{ij} \in \{0, 1\} \qquad\qquad \forall\, i, j \in N \tag{6}$$

$$Z_{ij} \geq 0 \qquad\qquad \forall\, i, j \in N \tag{7}$$

The objective function 1 aims the maximization of the coverage flow of the network. The constraint 2 ensures that only p-hubs are being allocated, while constraints 3 and 4 ensure that each spoke only has one hub associated and each node can be allocated only to one hub, respectively. In the end, constraints 5 are used to indicate the coverage flow percentage between the node pairs i and j. Note that parameter λ_{ij} is used to tighten constraints 5 and these assume values indicated by Eq. 8.

$$\lambda_{ij} = \max_{km}(a_{i,j}^{k,m}) \qquad\qquad \forall\, (i, j) \in N \tag{8}$$

However, it is possible to note that when the spoke j is connected to the hub m, $X_{jm} = 1$, constraints 5 are limited to $Z_{ij} \leq \sum_{k \in N} a_{i,j}^{k,m} X_{ik}$. In contrast, when $X_{jm} = 0$, the constraints take the configuration $Z_{ij} \leq \sum_{k \in N} a_{i,j}^{k,m} X_{ik} + \lambda_{ij}$. At last, the domain of variables X_{ij} and Z_{ij} are given by Eqs. 6 and 7, respectively.

3 Basic Variable Neighborhood Search Algorithm

In order to represent computationally the solution for the problem, a two-dimension array was used (two lines $\times$ n columns), referring to the amount of nodes in each instance. Table 1 shows how the solution from Fig. 1 could be represented.

Table 1. Solution representation of the example from Fig. 1

0	0	0	0	1	1	0	0	0	0
5	6	5	6	5	6	5	6	5	5

Note that Fig. 1 shows a feasible solution for the problem and can be obtained by the selection of 5 and 6 as initial hubs. Thus, to create an initial feasible solution, it is necessary to associate a set of p nodes to cover the flow service in the network.

To build the initial solution of the problem, two randomly nodes were chosen and the number of hubs was increased as each instance requests. The method of choice of the initial solution is justified because in the algorithm proposed to improve the solutions, all nodes will be tested as hubs. In the example in Table 1, nodes 5 and 6 were chosen as hubs to compose the initial solution, taking into account an instance with p equal to two. After the allocation of these nodes as hubs, all the others were allocated to these hubs considering the hub which presents the lowest cost.

When p is bigger than 2, after finding a solution with two hubs, the other hubs are incorporated to the current solution in such a way that maximizes the coverage of the network configuration. That is, while the total of hubs is lower than p, new hubs are added, one by one, having as selection criteria the hubs which enable the most significant enhancement in the coverage percentage of the network. Thus, once inserted p hubs, it is possible to obtain an initial solution for the problem.

In order to improve the solution quality, the metaheuristic known in the literature as Variable Neighborhood Search (VNS) was developed for the problem. Proposed by [7], this method aims to explore the set of solutions through systematic changes in some neighborhood structures with the aid of local searches procedures to find good solutions in the set of the possible ones. VNS pseudocode is presented in the Algorithm 1.

In Algorithm 1, the VNS input parameters are the current solution s (line line 1) and the number of iterations without any improvement ($Tmax$). In line line 2, the algorithm starts the variable s with the current solution obtained as a parameter. The loop in line line 3 is responsible to control how many times the procedure VNS will be applied, that is, how many iterations without any improvement will be executed.

In line line 4, a shake will be applied in the current solution. In line line 5, a local search procedure will be applied. In line line 6 it is verified if the solution s_0 is better than s and if so, solution s is updated by s_0 and t is reinitialized in 0, as proposed by lines line 7 and line 8, respectively. If the solution is not improved, t is increased in one unit (line 10) and the procedure returns the solution s.

A shake method in the current solution and a local search method were used in the development of VNS algorithm. The shake objectives to modify the present solution structure in order to escape from good local results and to search new

---

Algorithm 1. Basic Variable Neighborhood Search algorithm

1: **procedure** VNS(s, $Tmax$)
2:     $s_0 \leftarrow s$
3:     **while** $t \leq Tmax$ **do**
4:         $s_0 \leftarrow Shake(s_0)$
5:         $s_0 \leftarrow LocalSearch(s_0)$
6:         **if** $f(s_0) > f(s)$ **then**
7:             $s \leftarrow s_0$
8:             $t \leftarrow 0$
9:         **else**
10:             $t \leftarrow t + 1$
11:         **end if**
12:     **end while**
13:     **return** s
14: **end procedure**

---

options aiming to explore the new set of solutions and to reach better results. For this problem, the shake occurs as follows: two nodes, k and m (which are hubs), are chosen randomly and two nodes, i e j (which are spokes), also chosen randomly. The shake procedure occurs by the change in the hub-and-spoke configuration, that is, nodes that are spokes or hubs can be alternated changing its function or continue as before. The new hub-and-spoke configuration is given randomly considering 20% of probability to one of the following combinations happens:

$$First\ option = \begin{cases} i\ ,\ j \leftarrow hubs \\ m\ ,\ k \leftarrow spokes \end{cases}$$

$$Second\ option = \begin{cases} i\ ,\ k \leftarrow hubs \\ j\ ,\ m \leftarrow spokes \end{cases}$$

$$Third\ option = \begin{cases} i\ ,\ m \leftarrow hubs \\ j\ ,\ k \leftarrow spokes \end{cases}$$

$$Forth\ option = \begin{cases} j\ ,\ k \leftarrow hubs \\ i\ ,\ m \leftarrow spokes \end{cases}$$

$$Fifth\ option = \begin{cases} j\ ,\ m \leftarrow hubs \\ i\ ,\ k \leftarrow spokes \end{cases}$$

As stated before and considering Fig. 1 as example, the computational representation can be given by Table 1. From that, if the shake was applied on this solution, the hubs would be 5 and 6 ($m = 5$ and $k = 6$). In addition, two nodes would be randomly chosen as spokes, e.g., nodes 2 and 8 ($i = 2$ and $j = 8$). Since the nodes i, j, k and m were chosen, one from the five shake procedures are randomly applied. If the fourth option was chosen, e.g., the new hubs would be j and k, that is, nodes 8 and 6, and the new spokes i and m ($i = 2$ e $m = 5$) as presented in Table 2.

Table 2. Representation of the fourth pertubation option

Before									
0	0	0	0	1	1	0	0	0	0
5	6	5	6	5	6	5	6	5	5

After										
0	0	0	0	0	1	0	1	0	0
8	6	6	8	6	6	6	8	8	8

Regarding the local search proposed, a method in which all the nodes are tested as hubs was developed. For that, a spoke node and one hub are selected and the reallocation of the spoke as a new hub and the hub as a new spoke is done. From that, a reallocation procedure of the spokes aiming to maximize the coverage flow is also done. Table 3 represents how this modification in the solution structure would be done from the solution presented in Table 1.

Table 3. Representation of local search

Before									
0	0	0	0	**1**	**1**	0	0	0	0
5	6	5	6	5	6	5	6	5	5

After									
1	0	0	0	**0**	**1**	0	0	0	0
1	6	6	6	1	6	6	1	6	6

4   Computational Experiments

The VNS algorithm developed in this paper was implemented using the $C++$ programming language using a Windows 10 computer with an Intel Core i7-8550U 1.80 GHz processor with 4 cores and 8 GB of RAM memory. The commercial solver ILOG CPLEX 12.6 was used as the exact approach strategy of the instances evaluated, and it was implemented in AMPL language. Experiments made by [11], which are compared in this paper, were done in a computer with an Intel Xeon E7-2870 2.40 GHz processor with 10 cores, 20 threads and 512 GB of RAM memory.

It is important to highlight that the comparison presented in this paper is not totally appropriated because in numerical terms there are some differences between the machines used. In the TS algorithm from [11], the machine has a RAM memory of 512 GB and 2.40 Ghz of clock frequency, whereas the machine in which VNS was performed has 8 GB of RAM memory and 1.80 Ghz of clock frequency.

All the instances evaluated were executed once by the exact approach, that is, by CPLEX, and three times by VNS. For the VNS, the instances were executed with three different seeds and the number of iterations without any improvement was fixed in 60 based in some tests previously provided by the authors of this paper.

4.1   Instances Description

Two different sets of instances from the literature were evaluated in this paper: the Civil Aeronautics Board (CAB) and the Australian Post (AP).

Instances from CAB were firstly introduced by [9] and are commonly used in hub-and-spoke problems. CAB instances are arranged in relation to the number of nodes ($n = \{10, 15, 20, 25\}$), number of hubs to be fixed ($p = \{2, 3, 4, 5\}$) and the discount factor among the hubs ($\alpha = \{0.2, 0.4, 0.6, 0.8, 1.0\}$). Discounting factors χ and δ in all cases are set as 1, that is, there is no discount by the transport among the spokes once the problem does now allow connections among them. As an alternative comparison, the maximum cost values of route β were the same used by [11].

Regarding the AP instances, introduced by [3], instances with the number of nodes n equal to $\{10, 20, 25, 40, 50, 100\}$ were evaluated. In all of the cases, the discount factors α are fixed as 0.75, while χ and δ are fixed in 1. Again, in order to standardize the comparisons, β was defined to be the same proposed by [11] for each instance.

4.2   Results and Discussion

This section aims to explore the computational results obtained from the heuristic approaches VNS and TS, as well as the exact approach from the CPLEX

Table 4. Results of CAB instances with n set to 10

Instances				CPLEX			VNS	TS
n	p	α	β	optsol	cov(%)	t(s)	t(s)	t(s)
10	2	0.20	1425	994540	99.55	0.91	0.05	0.00
10	3	0.20	1117	999026	100.00	0.55	0.08	0.05
10	4	0.20	811	999026	100.00	0.36	0.08	0.08
10	5	0.20	736	991270	99.22	0.45	0.08	0.09
10	2	0.40	1627	994540	99.55	0.86	0.05	0.02
10	3	0.40	1185	990542	99.15	0.44	0.07	0.03
10	4	0.40	970	999026	100.00	0.36	0.08	0.08
10	5	0.40	863	999026	100.00	0.34	0.08	0.09
10	2	0.60	1671	987490	98.85	0.59	0.06	0.02
10	3	0.60	1387	984530	98.55	0.41	0.08	0.03
10	4	0.60	1148	999026	100.00	0.36	0.08	0.08
10	5	0.60	1079	999026	100.00	0.34	0.08	0.13
10	2	0.80	1744	999026	100.00	0.38	0.06	0.00
10	3	0.80	1589	999026	100.00	0.36	0.08	0.03
10	4	0.80	1457	999026	100.00	0.38	0.08	0.06
10	5	0.80	1413	999026	100.00	0.38	0.08	0.09
10	2	1.0	1839	984836	98.58	0.41	0.06	0.02
10	3	1.0	1791	999026	100.00	0.33	0.08	0.03
10	4	1.0	1770	999026	100.00	0.36	0.08	0.06
10	5	1.0	1766	999026	100.00	0.34	0.08	0.13
SGM						0.44	0.07	0.06

commercial solver. For easy reading we highlight the predefined parameters in all the result tables, to quote: the number of nodes (n), the number of hubs (p), the discount factor (α) and the maximum cost allowed β. Furthermore, Tables 4, 5, 6, 7 and 8 also present the optimal solution value obtained by CPLEX (*optsol*), the coverage percentage of the network flow (*cov*(%)) and the CPU computational time of the execution of the algorithms, in seconds ($t(s)$), required to find the optimal solutions, as well as the Shifted Geometric Mean [1] of the values (SGM).

The SGM was used to reduce the distortion between the average performance of the algorithms according to the dimension of the problems. We define 10 as the shifts values for the time ($t(s)$) as suggested by [1].

In cases in which one of the algorithms did not reach the optimal solution for the problems (Tables 5 and 7), the percentage distance from the obtained solution to the optimal is presented (*gap*(%)). Otherwise, when the values for the *gap* are equal to 0 (Tables 4 and 6) it means that the solutions presented by VNS, CPLEX and TS are optimal.

Table 5. Results of CAB instances with n set to 15

Instances				CPLEX			VNS		TS	
n	p	α	β	optsol	cov(%)	t(s)	gap(%)	t(s)	gap(%)	t(s)
15	2	0.20	2004	2358068	99.71	3.06	0.00	0.14	0.00	0.03
15	3	0.20	1638	2358068	99.71	1.88	0.00	0.26	0.00	0.19
15	4	0.20	1324	2364942	100.00	0.63	0.00	0.36	0.00	0.34
15	5	0.20	1149	2353712	99.53	0.88	0.00	0.48	0.00	0.58
15	2	0.40	2019	2364942	100.00	1.02	0.00	0.14	0.00	0.05
15	3	0.40	1741	2364942	100.00	0.66	0.00	0.26	0.00	0.19
15	4	0.40	1436	2364942	100.00	0.52	0.00	0.37	0.00	0.33
15	5	0.40	1287	2364942	100.00	0.41	0.00	0.43	0.00	0.59
15	2	0.60	2103	2364942	100.00	0.44	0.00	0.13	0.00	0.03
15	3	0.60	1844	2304218	97.43	1.20	0.00	0.29	0.00	0.19
15	4	0.60	1756	2364942	100.00	0.47	0.00	0.38	0.00	0.33
15	5	0.60	1560	2320434	98.12	0.44	0.00	0.45	0.00	0.59
15	2	0.80	2424	2364942	100.00	0.45	0.00	0.14	0.00	0.06
15	3	0.80	2165	2320434	98.12	0.69	0.00	0.25	0.00	0.19
15	4	0.80	2100	2364942	100.00	0.56	0.01	0.38	0.00	0.34
15	5	0.80	2080	2320434	98.12	0.50	0.00	0.45	0.00	0.59
15	2	1.00	2611	2364942	100.00	0.47	0.00	0.14	0.00	0.06
15	3	1.00	2610	2364942	100.00	0.55	0.00	0.25	0.00	0.16
15	4	1.00	2605	2364942	100.00	0.45	0.00	0.37	0.00	0.36
15	5	1.00	2600	2320434	98.12	0.52	0.00	0.48	0.00	0.56
SGM						0.77	0.00	0.31	0.00	0.29

Tables 4 and 5 show the results of the two smaller sets of CAB instances with nodes n set to 10 and 15, respectively. VNS and TS reached the optimal solution in almost all of the cases with the average time lower than the commercial solver CPLEX. For these two sets of smaller instances, VNS and TS were great alternatives to solve the problem, in both cases the optimal solution was obtained, as proved by CPLEX results, with a processing time lower than 1 second for most of the instances.

In a similar way, for larger instances with nodes n set to 20 and 25, the TS algorithm was more effective than VNS because it could reach the optimal solutions for all the cases, as shows in Tables 6 and 7. This scenario indicates that, although VNS is efficient to obtain good solutions in a reduced time, it can be improved in order to explore new solutions and escape from local optimal solutions, either by the addition of new neighborhood structures, or other type of initial solutions generation.

Table 6. Results of CAB instances with n set to 20

Instances				CPLEX			VNS	TS
n	p	α	β	optsol	cov(%)	t(s)	t(s)	t(s)
20	2	0.20	1851	5747720	99.88	8.95	0.37	0.19
20	3	0.20	1549	5743058	99.80	2.06	1.03	0.56
20	4	0.20	1356	5754594	100.00	4.70	1.29	1.08
20	5	0.20	1162	5722742	99.45	4.02	2.11	1.86
20	2	0.40	2067	5737094	99.70	3.27	0.36	0.17
20	3	0.40	1744	5739610	99.74	1.14	0.79	0.53
20	4	0.40	1473	5754594	100.00	1.05	1.31	1.11
20	5	0.40	1386	5754594	100.00	3.22	1.96	1.84
20	2	0.60	2255	5748824	99.90	3.17	0.38	0.16
20	3	0.60	1996	5719090	99.38	1.31	0.84	0.56
20	4	0.60	1835	5754594	100.00	0.67	1.38	1.09
20	5	0.60	1663	5754594	100.00	2.13	1.67	1.84
20	2	0.80	2493	5754594	100.00	0.89	0.38	0.17
20	3	0.80	2264	5754594	100.00	0.70	0.77	0.55
20	4	0.80	2154	5754594	100.00	0.84	1.31	1.11
20	5	0.80	2118	5752254	99.96	1.09	2.14	1.81
20	2	1.00	2611	5754594	100.00	1.11	0.42	0.19
20	3	1.00	2605	5754594	100.00	1.19	0.89	0.53
20	4	1.00	2601	5754594	100.00	1.11	1.38	1.08
20	5	1.00	2600	5710086	99.23	1.10	1.77	1.81
SGM						2.06	1.11	0.89

Table 7. Results of CAB instances with n set to 25

Instances				CPLEX			VNS		TS	
n	p	α	β	optsol	cov(%)	t(s)	gap(%)	t(s)	gap(%)	t(s)
25	2	0.20	2136	2358068	100.00	20.70	0.00	0.78	0.00	0.42
25	3	0.20	1913	2358068	99.93	49.42	0.00	1.72	0.00	1.34
25	4	0.20	1617	2364942	99.93	19.23	0.00	3.18	0.00	2.67
25	5	0.20	1346	2353712	100.00	4.81	0.00	4.53	0.00	4.52
25	2	0.40	2401	2364942	99.96	16.66	0.00	0.69	0.00	0.44
25	3	0.40	2099	2364942	100.00	13.55	0.04	1.68	0.00	1.48
25	4	0.40	1881	2364942	99.73	14.55	0.00	3.82	0.00	2.69
25	5	0.40	1597	2364942	99.84	2.73	0.00	4.36	0.00	4.52
25	2	0.60	2557	2364942	99.96	6.34	0.00	0.67	0.00	0.42
25	3	0.60	2336	2304218	99.96	4.70	0.00	1.66	0.00	1.36
25	4	0.60	2184	2364942	100.00	2.58	0.10	3.22	0.00	2.67
25	5	0.60	2002	2320434	99.81	6.28	0.00	5.51	0.00	4.50
25	2	0.80	2713	2364942	99.96	3.78	0.00	0.69	0.00	0.41
25	3	0.80	2552	2320434	99.96	2.92	0.00	1.71	0.00	1.34
25	4	0.80	2457	2364942	100.00	2.70	0.10	3.10	0.00	2.69
25	5	0.80	2307	2320434	99.42	3.38	0.00	5.29	0.00	4.48
25	2	1.00	2806	2364942	99.86	1.98	0.00	0.72	0.00	0.45
25	3	1.00	2762	2364942	100.00	1.84	0.00	1.91	0.00	1.30
25	4	1.00	2726	2364942	99.96	1.58	0.00	2.99	0.00	2.69
25	5	1.00	2725	2320434	99.96	1.72	0.00	4.41	0.00	4.53
SGM						7.08	0.01	2.54	0.00	2.15

For the set of AP instances, which contains more nodes, CPLEX was limited into 5 hours of execution. Results in bold present the best solution found for each instance. Besides, it is presented in Table 8 the percentage distance ($gap^b(\%)$) between the results obtain by VNS and TS. In some instances, CPLEX did not obtained any solution because the machine in which the tests were executed does not have memory enough to solve it and because of that the results were represented by a symbol (–) in Table 8. Note that for the majority of instances the algorithms found the same solutions for the problems (Table 8). It is highlighted that VNS did not reach the same solutions when compared with the TS in some of the large-scale instances, although the $gap^b(\%)$ is very tiny.

In contrast, the computational effort spent by VNS to reach solutions equal to TS or with a little bit difference is sufficiently slower, which indicates that the solutions are equivalents, but VNS also had two best solutions when compared with TS for two large instances ($n = 100$).

Table 8. Results of AP instances

Instances				CPLEX			VNS			TS		
n	p	α	β	optsol	cov(%)	t(s)	solVNS	gap^b(%)	t(s)	solTS	gap^b(%)	t(s)
10	2	0.75	40383	**3978.915**	100.00	0.36	**3978.915**	0.00	0.04	**3978.915**	0.00	0.02
10	3	0.75	34772	**3937.305**	98.95	0.47	**3937.305**	0.00	0.05	**3937.305**	0.00	0.05
10	4	0.75	32574	**3954.533**	99.39	0.44	**3954.533**	0.00	0.09	**3954.533**	0.00	0.10
10	5	0.75	32531	**3954.533**	99.39	0.63	**3954.533**	0.00	0.08	**3954.533**	0.00	0.13
20	2	0.75	45954	**3973.210**	99.86	1.33	**3973.210**	0.00	0.30	**3973.210**	0.00	0.20
20	3	0.75	43400	**3973.198**	99.86	2.13	**3973.198**	0.00	0.69	**3973.198**	0.00	0.60
20	4	0.75	38607	**3974.269**	99.89	0.61	**3974.269**	0.00	1.20	**3974.269**	0.00	1.21
20	5	0.75	37868	**3973.198**	99.86	0.81	**3973.198**	0.00	1.63	**3973.198**	0.00	2.01
25	2	0.75	53207	**3976.570**	99.94	2.94	**3976.570**	0.00	0.72	**3976.570**	0.00	0.46
25	3	0.75	46608	**3972.507**	99.84	2.80	**3972.507**	0.00	1.71	**3972.507**	0.00	1.47
25	4	0.75	45552	**3976.681**	99.94	2.20	**3976.681**	0.00	2.93	**3976.681**	0.00	2.95
25	5	0.75	45552	**3976.681**	99.94	1.98	**3976.681**	0.00	4.31	**3976.681**	0.00	5.06
40	2	0.75	61683	**3978.915**	100.00	128.47	**3978.915**	0.00	4.24	**3978.915**	0.00	3.29
40	3	0.75	58193	**3978.915**	100.00	8.00	**3978.915**	0.02	11.92	**3978.915**	0.00	10.13
40	4	0.75	52265	**3977.276**	99.96	185.59	**3977.276**	0.00	22.17	**3977.276**	0.00	20.82
40	5	0.75	49741	**3977.966**	99.98	21.78	**3977.966**	0.02	33.91	**3977.966**	0.00	35.62
50	2	0.75	65523	–	–	–	**3978.688**	0.00	10.66	**3978.688**	0.00	7.90
50	3	0.75	60132	–	–	–	**3978.415**	0.00	33.65	**3978.415**	0.00	24.16
50	4	0.75	52906	**3978.915**	100.00	15.73	3978.032	0.02	63.00	**3978.915**	0.00	49.05
50	5	0.75	50708	**3978.915**	100.00	12.36	3977.623	0.03	97.66	**3978.915**	0.00	82.87
100	2	0.75	65915	–	–	–	**3978.851**	0.00	169.31	3978.688	0.00	7.90
100	3	0.75	60659	–	–	–	**3978.590**	0.00	539.99	3978.415	0.00	24.16
100	4	0.75	56125	–	–	–	3978.620	0.00	981.37	**3978.915**	0.00	49.05
100	5	0.75	54243	–	–	–	3978.725	0.00	1550.64	**3978.915**	0.00	82.87
200	2	0.75	68232	–	–	–	3978.818	0.00	3457.71	**3978.915**	0.00	2749.00
200	3	0.75	64237	–	–	–	3978.833	0.00	11455.46	**3978.915**	0.00	7677.75
200	4	0.75	59999	–	–	–	3978.833	0.00	15852.32	**3978.915**	0.00	15987.99
200	5	0.75	58562	–	–	–	3978.833	0.00	42003.27	**3978.915**	0.00	26731.48
SGM								0.02	65.83		0.00	48.21

5   Conclusions and Remarks

This paper presented a study of the problem known in the literature as Uncapacitated Single Allocation p-hub Maximal Covering Problem (USApHMCP). This problem consists in selecting p-hubs in a network design, selecting nodes to be hubs in such a way to maximize the coverage of the network, considering that the remote nodes (which are not hubs) must be allocated to a single hub. From this, it was proposed an algorithm based on the Basic Variable Neighborhood Search metaheuristic to solve USApHMCP. Two different set of test instances from the literature, the Civil Aeronautics Board (CAB) and the Australian post (AP), are evaluated. The computational performance of our VNS and a Tabu Search (TS) proposed by [11] are compared in solution quality.

Computational results have shown that VNS and TS are good alternatives to solve the problem. Both reached optimal solutions, as proved by CPLEX, to the majority of the instances considered. For the CAB instances, the TS found the optimal solution in all of the cases. Even though the algorithms were executed in machines with different capacity of processing, it is possible to realize that VNS is a great alternative to solve the problem considering the solutions presented. In an opposite way, VNS did not reach the optimal solution in about 5% (4 of 80) of the CAB instances with the SGM $gap(\%)$ lower than 0.10% in the worst case. In relation to the large-scale AP instances, that is, the instances with the highest number of nodes, it is verified that both VNS and TS reached good solutions in a timely manner, yet, VNS reached two results that are better when compared with those presented by TS.

At last, as the next stage of the research, new strategies to generate initial solutions and neighborhood structures, as well as other metaheuristics, will be implemented aiming to improve the solutions presented until then.

Acknowledgements. The authors acknowledge the UFOP, Fapemig, CAPES and CNPq for supporting this research.

References

1. Achterberg, T.: Constraint integer programming. Ph.D. thesis, Technische Universität Berlin, Fakultä II - Mathematik und Naturwissenschaften (2007)
2. Campbell, J.: Integer programming formulations of discrete hub location problems. Eur. J. Oper. Res. **72**(2), 387–405 (1994)
3. Ernst, T., Krishnamoorthy, M.: Efficient algorithms for the uncapacitated single allocation p-hub median problem. Locat. Sci. **4**(3), 139–154 (1996)
4. Gendreau, M., Potvin, J.-Y.: Tabu search. In: Gendreau, M., Potvin, J.-Y. (eds.) Handbook of Metaheuristics. ISORMS, vol. 272, pp. 37–55. Springer, Cham (2019). https://doi.org/10.1007/978-3-319-91086-4_2
5. Janković, O., Mišković, S., Stanimirović, Z., Todosijević, R.: Novel formulations and VNS-based heuristics for single and multiple allocation p-hub maximal covering problems. Ann. Oper. Res. **259**(1–2), 191–216 (2017)
6. Kara, B.Y., Tansel, B.C.: The single-assignment hub covering problem: models and linearizations. J. Oper. Res. Soc. **54**(1), 59–64 (2003)
7. Mladenović, N., Hansen, P.: Variable neighborhood search. Comput. Oper. Res. **24**(11), 1097–1100 (1997)
8. O'Kelly, M.E.: Activity levels at hub facilities in interacting networks. Geogr. Anal. **18**(4), 343–356 (1986)
9. O'Kelly, M.E.: A quadratic integer program for the location of interacting hub facilities. Eur. J. Oper. Res. **32**(3), 393–404 (1987)
10. Peker, M., Kara, B.Y.: The P-Hub maximal covering problem and extensions for gradual decay functions. Omega **54**, 158–172 (2015)
11. Silva, M.R., Cunha, C.B.: A tabu search heuristic for the uncapacitated single allocation p-hub maximal covering problem. Eur. J. Oper. Res. **262**(3), 954–965 (2017)

A Variable Neighborhood Search Algorithmic Approach for Estimating MDHMM Parameters and Application in Credit Risk Evaluation for Online Peer-to-Peer (P2P) Lending

Monir El Annas[✉], Mohamed Ouzineb, and Badreddine Benyacoub

Institut National de Statistique et d'Economie Appliquée, Rabat, Morocco
elannas.mounir@gmail.com, ouzineb.insea@gmail.com, benyacoubb@gmail.com

Abstract. Online peer-to-peer (P2P) lending is a new financing channel on which lenders are matched with borrowers using internet platform. Borrowers can get financing more easily, but it means higher credit risk to lenders, making credit scoring models a key tool for lending P2P platforms. The goal is to estimate the level risk (being good or bad borrower), from the collected informations of each applicant. One of the classification approaches is Multi dimensional Hidden Markov Model (MDHMM). The MDHMM parameters are usually estimated using Baum-Welch algorithm (BW). However, the Baum-Welch algorithm tends to arrive at local optimal points. In this paper a strategy called Variable neighborhood search (VNS), is proposed to addresses this problem. The hybrid model in which VNS algorithm is coupled with Baum-Welch algorithm for parameter estimation of MDHMM, is applied in credit scoring domain, using real peer-to-peer lending data. The experiments results show the performance efficiency of our model in comparison with classical and alternative machine learning models for credit scoring.

Keywords: Multi dimensional Hidden Markov Model · Baum-Welch Algorithm · VNS algorithm · Credit scoring · P2P lending

1 Introduction

After the recent world financial crisis, more attention was given from banks and financial institutions to credit risk, since it can cause great cost losses to owners, managers, workers, lenders, clients, community and government. Therefore, it is very important to predict bankruptcy and decide whether to grant credit to new applicants or not. One of the primary tools used by banks is credit scoring. The problem in the credit scoring generally, is presented as a classification task where the applicant may be assigned to a class target (good or bad) based on their characteristic such Age, Salary and Housing... The goal is to estimate the level risk

© Springer Nature Switzerland AG 2020
R. Benmansour et al. (Eds.): ICVNS 2019, LNCS 12010, pp. 139–151, 2020.
https://doi.org/10.1007/978-3-030-44932-2_10

(being good or bad), from the collected information of each applicant. One o the classification approaches is by modeling the risk level using HMM models with two hidden states where each state represent a different risk category. MDHMM can be employed to modulate the behavior of borrowers and estimate the risk being good or bad using multiple observation sequences where each sequence correspond to an input variable (characteristic). Those HMM models are usually trained by Baum-Welch Algorithm. However, Baum-Welch algorithm tends to arrive at local optimal points. A possible way to escape Baum-Welch algorithm (BWA) local optimum is by using VNS algorithm. VNS is based on the application of local search by systematically changing the neighborhood during the search.

The rest of this paper is structured as follows. Section 2 is a comprehensive literature review concerning credit scoring studies in p2p lending and metaheuristics used for hidden Markov models training. Section 3 is an introduction to HMM, MDHMM, BWA, and VNS associated notations and algorithms. Section 4 is a description of the data used in this paper and explains the experiment setup and results. Finally, Sect. 5 concludes and discusses future work.

2   Literature Review

2.1   Credit Scoring in P2P Lending

Peer-to-peer (P2P) lending platforms (also known as social lending) are new financial intermediary between borrowers and lenders are, emerging rapidly worldwide. For example, loans the biggest P2P lending platform in USA between June 2007 and June 2018, accepted 2,004,090 loans, and rejected 22,469,074 loans, issued loans are roughly 8% of all loan requests on their website [16,17]. Given the critical role of credit scoring in P2P lending, various studies focused on the specific P2P lending domain. These models are categorized into statistical models and AI-based models. LR is one of the most popular statistical models in P2P lending mainly due to its acceptable performances and interpretability. AI-based methods, such as SVM, HMM, Random forest and Neural network have been applied in credit scoring of P2P lending due to their superior predictability [1–6].

2.2   Metaheuristics for Hidden Markov Models Training

Seven main kinds of generic metaheuristics have been adapted to tackle the issue of HMMs training: the simulated annealing (SA), the tabu search (TS), genetic algorithms (GA), the population based incremental learning (PBIL), the API algorithm and the particle swarm optimization (PSO) [26–31]. All those adaptations do not try to maximize the same criterion but the maximum likelihood criterion is the most often used.

3 Methodology

3.1 Hidden Markov Model

Elements of a Hidden Markov Model
An HMM model is characterized by the following elements [7]:

- S_t The random variable representing the state at time t, where $0 \le S_t \le N-1$ and $0 \le t \le T - 1$, N is the number of states in the model, and T being the length of the observation sequence. Let $S = (S_0, S_1, \ldots, S_{T-1})$ be the states sequence.
- O_t The random variables representing the observation at time t, where $0 \le O_t \le M$ and $0 \le t \le T$, where M is the number of observation symbols, and T the length of the observation sequence. Let $O = (O_0, O_1, \ldots, O_{T-1})$ be the observation sequence.
  In this paper we consider the values of O_t and the values of S_t to be discrete.
- $A = \{a_{ij}\}$ the state transition probabilities matrix, where $A \in R^{N \times N}$ and $a_{ij} = P(S_{t+1} = j | S_t = i)$
- $B = \{b_i(k)\}$ the observation probability matrix, where where $B \in R^{N \times M}$ and $b_i(k) = P(O_t = k | S_t = i)$
- $\pi = \{\pi_i\}$ be the initial probability vector, where $\pi \in R^N$.
  and $\pi_i = P(S_t = i)$
- we denote an HMM as a triplet $\lambda = (\pi, A, B)$

Training with Baum-Welch algorithm
Baum-Welch is a learning algorithm [10] that is based on the principles of Expectation Maximization (EM) [11] to find the optimum model parameter λ that maximizes $P(O|\lambda)$ we first define the following probabilities:

- $\alpha_t(i) = P(O_0, O_2, \ldots, O_t, S_t = i | \lambda)$
- $\beta_t(i) = P(O_{t+1}, O_{t+2}, \ldots, O_{T-1} | S_t = i, \lambda)$
- $\gamma_t(i) = P(S_t = i | O, \lambda)$
- $\zeta_t(i, j) = P(S_t = i, S_{t+1} = j | O, \lambda)$

Then given a random initial conditions for λ, (it can also be set using prior information about the parameters if it is available).:
for $i = 0, 1 \ldots, N - 1$ and $t = 1, 2 \ldots, T - 1$

- $\alpha_0(i) = \pi_i b_i(O_0)$ and $\alpha_t(i) = [\sum_{j=0}^{N-1} \alpha_{t-1}(j) a_{ji}] b_i(O_t)$
  for $i = 0, 1 \ldots, N - 1$ and $t = T - 2, T - 3 \ldots, 0$
- $\beta_{T-1}(i) = 1$ and $\beta_t(i) = \sum_{j=0}^{N-1} a_{ij} b_j(O_{t+1}) \beta_{t+1}(j)$
  for $i, j \in 0, \ldots, N - 1$ and $t = 0, 2 \ldots, T - 2$
  - $\zeta_t(i, j) = \frac{\alpha_t(i) a_{ij} b_j(O_{t+1}) \beta_{t+1}(j)}{\sum_{k=0}^{N-1} \alpha_{T-1}(k)}$
    for $i = 0, 1 \ldots, N - 1$ and $t = 0, 2 \ldots, T - 2$
  - $\gamma_t(i) = \frac{\alpha_t(i) \beta_t(i)}{\sum_{k=0}^{N-1} \alpha_{T-1}(k)} = \sum_{j=0}^{N-1} \zeta_t(i, j)$

The parameter reestimation formulas, are described by the following expressions:

for $i, j \in 0, \ldots, N-1$ and $k = 0, 1 \ldots, M-1$

- $\widehat{\pi}_i = \gamma_0(i)$

- $\widehat{a_{ij}} = \dfrac{\sum_{t=0}^{T-2} \zeta_t(i,j)}{\sum_{t=0}^{T-2} \gamma_t(i)}$

- $\widehat{b_i(k)} = \dfrac{\sum_{t=0, O_t=k}^{T-1} \gamma_t(i)}{\sum_{t=0}^{T-1} \gamma_t(i)}$

Given a sequence of observed symbols O the probability that the sequence was generated by HMM with parameters λ is: $P(O|\lambda) = \sum_{j=0}^{N-1} \alpha_{T-1}(j)$

The parameter λ will be estimated iteratively by Baum-welch procedure as follow:

1. Start with a random guess of parameters $\lambda = \lambda_0$ for the model
2. Estimate $\alpha_t(i)$ and $\beta_t(i)$, $\zeta_t(i,j)$, $\gamma_t(i)$ using the observation sequence O and the parameters of λ as shown in the equations
3. Update λ using equations and calculate $P(O|\lambda)$
4. Repeat steps 2 and 3 until a convergence to a stationary point of the likelihood. (i.e: $P(O|\lambda)$ is a local maximum).

3.2   Multi Dimensional HMM

MDHMM can be considered as an extension of hidden Markov model. The difference is in the number of observed variables. HMM defines only a single observed variable, whereas the MDHMM supports multiple observed variables with one common hidden sequence. In this paper we will assumes independence between the different dimensions of the input data.

Elements of MDHMM

The parameters of the multi-dimensional HMM can be defined with the same assumptions for random variable S_t where N is the number of states and T is the length of the observation sequence $S = (S_0, S_1, \ldots, S_{T-1})$. Note also that the matrix A corresponding to state transition probabilities and π the initial probability vector have the same structure as defined in the Sect. 3.1. Then, $P(S_t|S_{t-1})$ is represented by the matrix A. For calculating observation probability matrix, we suppose that the situation depicted in single observation sequence model will be extended to a multiple observation sequences as follows [12–14].

- O_t The random variables representing the observation at time t, where $O_t = \{O_t^0, O_t^1, \ldots, O_t^{M-1}\}$ and $0 \leq t \leq T-1$, M is the number of observed variables, and T is the length of the observation sequence.
- $O = (O_0, O_1, \ldots, O_{T-1})$ be the observation sequence from the M variables.
- $O^m = (O_0^m, O_1^m, \ldots, O_{T-1}^m)$ the m-th sequence of observation In this paper we consider the values of observed symbols O_t^m and the values of S_t to be discrete.

- $B^m = \{b_i^m(k)\}$ the m-th observation probability matrix, where $B^m \in R^{N \times L_m}$ and $b_i^m(k) = P(O_t^m = o_k^m | S_t = i)$ where o_k^m is a possible value of the random variable of the observation O_t^m. L_m being the number of observations symbols of m-th observation sequence.
- We denote an MDHMM as a triplet $\lambda = (\pi, A, B^{0:M-1})$
- An MDHMM has the joint probability function:
  $P(O_t, S_t) = P(S_{t-1})P(S_t|S_{t-1}) \prod_{m=0}^{M-1} P(O_t^m|S_t)$
- Each $P(O_t^m|S_t)$ is represented by the matrix B^m.

The learning approach is similar to the one described in the case of HMM. The only difference is that there are multiple emission functions B^m.

The parameter learning process can be performed by means of the Baum-Welch algorithm extended for MDHMM based on the probabilities $\alpha_t(i)$, $\beta_t(i)$, $\gamma_t(i)$ and $\gamma_t(i,j)$. Computed as follows:

- $\alpha_0(i) = \pi_i \prod_{m=0}^{M-1} b^m(O_0^m)$
- $\alpha_t(i) = [\sum_{j=0}^{N-1} \alpha_{t-1}(j)P(S_t = i, S_{t-1} = j, \lambda)] \prod_{m=0}^{M-1} P(O_t^m|S_t = i, \lambda)$
  $= [\sum_{j=0}^{N-1} \alpha_{t-1}(j)a_{ji}] \prod_{m=0}^{M-1} b^m(O_t^m)$
- $\beta_t(i) = \sum_{j=0}^{N-1} P(S_{t+1} = i, S_t = j, \lambda) \prod_{m=0}^{M-1} P(O_{t+1}^m|S_{t+1} = i, \lambda)\beta_{t+1}(i)$
  $= [\sum_{j=0}^{N-1} a_{ij}\beta_{t+1}(i)] \prod_{m=0}^{M-1} b^m(O_{t+1}^m)$
- $\beta_{T-1}(i) = 1$
- $\zeta_t(i,j) = \frac{\alpha_t(i)\beta_{t+1}(j)a_{ij} \prod_{m=0}^{M-1} b^m(O_{t+1}^m)}{\sum_{j=0}^{N-1} \alpha_{T-1}(j)}$
- $\gamma_t(i) = \frac{\alpha_t(i)\beta_t(i)}{\sum_{j=0}^{N-1} \alpha_{T-1}(j)} = \sum_{j=0}^{N-1} \zeta_t(i,j)$

The variables $\alpha_t(i), \beta_t(i), \zeta_t(i,j), \gamma_t(i)$ are calculated for each training sequence O^m and then, re-estimation is performed on the accumulated values. To generate the new better model $\widehat{\lambda}$, the parameter λ can be updated as follows:

- $\widehat{\pi_i} = \gamma_0(i)$
- $\widehat{a_{ij}} = \frac{\sum_{t=0}^{T-2} \zeta_t(i,j)}{\sum_{t=0}^{T-2} \gamma_t(i)}$
- $\widehat{b_i^m(k)} = \frac{\sum_{t=0/O_t^m=o_k^m}^{T-1} \gamma_t(i)}{\sum_{t=0}^{T-1} \gamma_t(i)}$

The iterative process described above is performed until a convergence to a stationary point of the likelihood.

3.3   VNS Algorithm

VNS [19–24] is based on a simple principle: the systematic change of neighborhoods within the search. It explores increasingly distant neighborhoods of the current solution, jumping from this solution to a new one if and only if an improvement has been made. Working in this way, favorable characteristics of the current solution will be often kept and used to obtain promising solutions.

Moreover, in order to obtain local optima, a local search routine is repeatedly applied to these neighboring solutions. VNS algorithm performs as follows:

Firstly, a finite set of neighborhood structures N_j, $(j = 1, \ldots, n)$ is defined. Next, an initial solution S is randomly generated or using a constructive heuristic. At each iteration, the neighborhood index (j) is initialized to 1. VNS's loop is composed of three steps: shaking, local-search and move. In the shaking step, a solution S' in the jth neighborhood of the incumbent solution S is randomly generated. Then, a local-search procedure is applied to the shaking's output S'. The local search's output is denoted $\widehat{S}$. If $\widehat{S}$ is better than S, $\widehat{S}$ replaces S from which the search continues with $j = 1$. Otherwise, j is incremented and a new shaking step starts using the $(j+1)^{th}$ neighborhood structure. The process continues until a stopping condition is met. The stopping condition may be the maximum number of iterations allowed and/or the maximum computational time allowed. The pseudocode of a VNS algorithm is given by Algorithm 1.

Algorithm 1. VNS algorithm

1: procedure VNS(S)
2: $\quad S \leftarrow$ InitialSolution ()
3: $\quad$ define a set of neighborhood structures N_j, $(j = 1, \ldots, n)$
4: $\quad$ **while** stopping condition is not met **do**
5: $\qquad j \leftarrow 1$
6: $\qquad$ **while** $j < n$ **do**
7: $\qquad\qquad S' \leftarrow$ Shaking $(N_j(S))$
8: $\qquad\qquad \widehat{S} \leftarrow$ LocalSearch (S')
9: $\qquad\qquad$ **if** $\widehat{S}$ is better then S **then**
10: $\qquad\qquad\qquad S \leftarrow \widehat{S}$
11: $\qquad\qquad\qquad j \leftarrow 1$
12: $\qquad\qquad$ **else**
13: $\qquad\qquad\qquad j \leftarrow j + 1$
14: $\qquad\qquad$ **end if**
15: $\qquad$ **end while**
16: $\quad$ **end while**
17: $\quad$ return S
18: end procedure

3.4   Estimating MDHMM Parameters Using BW and VNS for the Credit Scoring Problem

The problem in the credit scoring generally, is presented as a classification task where the applicant may be assigned to a class target (good or bad) based on their characteristic such Age, Salary and Housing...... Our goal is to estimate the level risk (being good or bad), from the collected information of each applicant. We take risk level to be represented by two hidden states where each state represent a different risk category. In this work, MDHMM can be employed to modulate the behavior of borrowers and estimate the risk being good or bad using multiple observation sequences where each sequence correspond to an input variable (characteristic). In our setup, a sequence of observable symbols preprocessed into a canonical form for different features.

We assumed that the observable symbols were dependent on a set of 2 hidden states to be $S_0 =$ good applicant, $S_1 =$ bad applicant. For each feature $m, m = 1, 2, \ldots, M$ we calculate B^m and we built a MDHMM model $\lambda = (\pi, A, B)$ where $B = B^{0:M-1}$.

The observed sequences $O^m = (O_0^m, O_1^m, \ldots, O_{T-1}^m), m = 1, 2, \ldots, M$ will be used to estimate MDHMM parameters as described in Sect. 3.2. Therefore, we can perform the iterative calculation of the $\alpha_t(i), \beta_t(i), \zeta_t(i, j), \gamma_t(i)$. Given an observation vector of information applicant noted by $O_t = \{O_t^0, O_t^1, \ldots, O_t^{M-1}\}$. The developed MDHMM model returns the most likely state for each applicant using the probability $\gamma_t(i)$.

From the definition of $\gamma_t(i) = P(S_t = i|O, \lambda)$ it follows that the candidate state at time t is the state S_i for which $\gamma_t(i)$ is maximum, where the maximum is taken over the index i. The iterative procedure of Baum-Welch is used to improve the estimated parameters of HMM and to adjust the parameters to observed sequence. However, it has a limited convergence to a local optimum. VNS could overcome the drawbacks of HMM using the neighborhood of the current solution to search existing other best solutions iteratively. The hybrid approach can be exploited to improve the performance of the developed model for credit scoring.

For single HMM we represent parameter $\lambda = (\pi, A, B)$ of feature i as a vector W_i. For example, a HMM with two hidden states and five observable variables would be represented as a string W_i where $|W_i| = |\pi| + |A| + |B| = 16$. Each $W_{i,j}$ could have any values. However, only positive values would be used. Normalization would ensure that the HMM has well-formed probability entries. Then, we extend the representation of parameters for all input variables.

For generating the initial solution for the proposed VNS algorithm, the Baum-Welch algorithm is used to construct a set of feasible solutions and then select the best solution as the starting point for the proposed VNS algorithm. In the proposed VNS algorithm, three neighborhood structures, $N_1()$, $N_2()$ and $N_3()$, are used. Let S $= \lambda_0 = (\pi, A, B)$; be a feasible solution $N_1(S)$: the set of solutions obtained by adding 1 for each $W_{i,j}$ and keep the remaining and do a normalization. $N_2(S)$: the set of solutions obtained by adding 2 for each $W_{i,j}$ and keep the remaining and do a normalization. $N_3(S)$: the set of solutions obtained by adding 3 for each $W_{i,j}$ and keep the remaining and do a normalization. Each iteration on W_i elements represent one plausible set of HMM parameters for the feature i.

In the proposed model two phases are considered to present the whole process of the hybrid model. In the first phase, we choose initial parameters $\lambda^0 = (\pi^0, A^0, B^0)$, using the Baum-Welch algorithm on the training set, we re-estimate parameters iteratively $\lambda = \lambda^0$ until to find the optimal model $\lambda = \lambda^* = (\pi^*, A^*, B^*)$. Therefore, the optimal model obtained will be evaluated on the testing test. In the second phase, the VNS process of the hybrid model is start from the parameters of the optimal model generated from Baum-Welch procedure. The optimizing process of the hybrid model was first presented by Baum-Welch which consist to find parameters that maximize the likelihood of observation sequences. Then, the VNS method was start from the HMM parameters generated by Baum-Welch procedure where the obtained optimal model will be used as initial solution. In order to maximize the accuracy of credit scoring model, a set of new parameters given by neighborhood of current solution are constructed and tested. VNS process continues to generate neighboring solutions (candidate credit scoring models) until reach convergence. This process is shown in the following figure (Fig. 1).

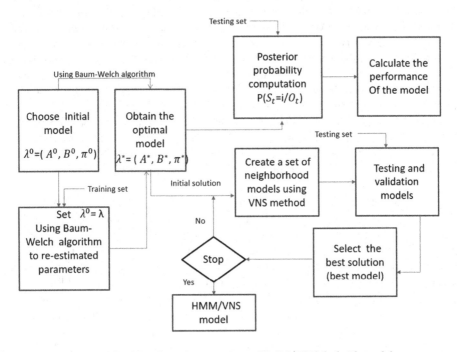

Fig. 1. The flow chart for creating a HMM/VNS hybrid model

4 Experimental Setup

4.1 Data Description

The P2P dataset we used is from Lending Clubs dataset [17]. Excluding records containing obvious errors and the characteristics, with missing information, and by keeping accepted records, with both good/bad statuses observed, we got a dataset consisting of 799,443 issued loans including 158,592 defaults, with 5 attributes, which 3 numerical, and 2 categorical: loan amount, FICO score, DTI ratio, address state, employment length.

We will use a balanced dataset of the Lending Clubs dataset created by the SMOTE technique (Synthetic Minority Over-sampling Technique), it's an over-sampling method, it creates synthetic samples of the minority class. Hence making the minority class equal to the majority class. SMOTE does this by selecting similar records and altering that record one column at a time by a random amount within the difference to the neighbouring records.

4.2   K Fold Corss Validation

To minimize the impact of data dependency and improve the reliability of the estimates, k-fold cross validation is used to create random partitions of the data sets. The procedure of k-fold cross-validation is as follow:

1. The data set is split into k mutually folds of nearly equal size.
2. Choose the first subset for testing set and the $k-1$ remainder for training set.
3. Build the model on the training set.
4. Evaluate the model on the testing set by calculating the evaluation metrics.
5. Alternately choose the following subset for testing set and the $k-1$ remainder for training set.
6. The structure of the model is then trained k times each time using $k-1$ subsets (training set) for training and the performance of the model is evaluated $k-1$ on the remaining subset (testing set).
7. The predictive power of classifier is obtained by averaging the k validation fold estimates found during the k runs of the cross validation process.

The common values for k are sometimes 5 and 10. The Cross validation method is used in this work to assess the performance of classification techniques and we choose 10 as value for k for our experiments evaluation method. This approach can be computationally expensive, but does not waste too much data (as it is the case when fixing an arbitrary test set), and lower the variance of the estimate.

The whole building process of model can be presented into three phases, which are shown in Fig. 2. First, the cross validation procedure is applied to split the dataset into two different subsets. Secondly, the VNS-MDHMM model is trained by the training set and the multiple emission functions B^m parameters are estimated. Then the estimated B^m are employed to classify the observations given in testing set and finally evaluate the performance of the model.

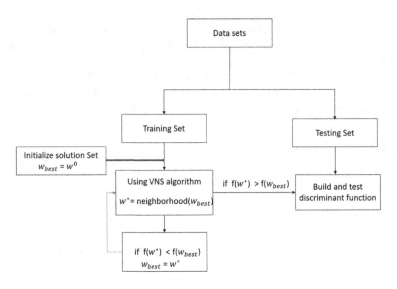

Fig. 2. Block diagram of the HMM/VNS hybrid model

4.3   Evaluation Metrics

There are many metrics that can be used to measure the performance of a classifier different fields have different preferences for specific metrics due to different goals. In this paper the performance of the models used was first measured in terms of average accuracy, precision, recall, and F1-score. The following equations show the process to calculate the accuracy, precision, recalls and F1-score:

$$Accuracy = \frac{TP+TN}{TP+TN+FP+FN} \qquad Precision = \frac{TP}{TP+FP}$$
$$Recall = \frac{TP}{TP+FN} \qquad F1 = \frac{2(Precision * Recall)}{Precision + Recall} = \frac{2TP}{2TP+FP+FN}$$

Where true positive (TP) is the number of instances that actually belong to the good group that were correctly classified as good by the classifier, true negative (TN) is the number of instances that belong to the bad group and correctly classified as bad, false positive (FP) is the number of instances that are of the bad group but mistakenly classified as good, and, finally, false negative (FN) is the number of instances that are actually of good but incorrectly classified as bad.

4.4   Computation

We conduct the experimental analysis for the proposed method and compared the results with other models, using Scikit-learn an open source Python library that provides a range of machine learning algorithm, and we used GridSearchCV for parameters setting and combinations.

4.5   Experiments Results

Table 1 compare the accuracy, accuracy, precision, recall, and F1-score for methods namely SGDC, RandomForest, SVM, KNN, MDHMM, and VNS-MDHMM performed with 10 fold cross validation.

Table 1. Comparison performance of the models over P2P lending dataset

	Measures of performance			
	Accuracy	Precision	Recall	F1-score
SGDC	53.06	31.35	59.97	40.00
RandomForest	**84.15**	**85.68**	79.07	79.11
SVM	51.51	59.54	65.93	46.45
KNN	53.44	50.73	81.29	62.58
MDHMM	61.15	72.53	58.65	64.85
VNS-MDHMM	61.43	61.31	**83.72**	70.78

The results show that, the best accuracy is obtained by RandomForest 84.15% followed by VNS-MDHMM with 61.43% improving the accuracy of MDHMM which is 61.15% followed by SVM 51.51%, KNN with 53.44% and SGDC 53.06%. But the Recall for VNS-MDHMM is highest than other classification models with 83.72% followed by KNN with 81.29% and RandomForest with 79.07% followed by the rest of the models. A key advantage of the proposed model over the existing models is that, we can build a set of models based on iterative procedure. But since the Baum-Welch algorithm and the VNS algorithm are local iterative methods, the resulting MDHMM performance depend heavily on the initial model.

5   Conclusion and Future Research

In this paper, a simple version of the VNS algorithm is employed to search out the optimal parameter structure of MDHMM for credit scoring in P2P lending. Experimental results show that the classification performance of the proposed have a high prediction rate and comparable performance to other widely used methods in credit scoring. As future work, it would be interesting to build a parallel implementation of our model [25], especially with the grow of big data sources like mobile phone data and social network data for credit scoring applications [18].

References

1. Lessmann, S., Baesens, B., Thomas, L.C., Seow, H.-V.: Benchmarking state-of-the-art classification algorithms for credit scoring: an update of research. Eur. J. Oper. Res. **247**(1), 124–136 (2015)

2. Harris, T.: Credit scoring using the clustered support vector machine. Expert Syst. Appl. **42**(2), 741–750 (2015)
3. Xia, Y., Liu, C., Li, Y., Liu, N.: A boosted decision tree approach using Bayesian hyper-parameter optimization for credit scoring. Expert Syst. Appl. **78**, 225–241 (2017)
4. Byanjankar, A., Heikkilä, M., Mezei, J.: Predicting credit risk in peer-to-peer lending: a neural network approach. In: Proceedings of the IEEE Symposium Series on Computational Intelligence, pp. 719–725, December 2015
5. Monir, E.A., Ouzineb, M., Benyacoub, B.: Multi dimensional Hidden markov model for credit scoring systems in Peer-To-Peer (P2P) lending. In: Farhaoui, Y. (ed.) BDNT 2019. LNNS, vol. 81, pp. 73–83. Springer, Cham (2020). https://doi.org/ 10.1007/978-3-030-23672-4_7
6. Ge-Er, T., Chang-Zheng, H., Jin, X., Xiao-Yi, J.: Customer credit scoring based on HMM/GMDH hybrid model. Knowl. Inf. Syst. **36**(3), 731–747 (2013). https:// doi.org/10.1007/s10115-012-0572-z
7. Rabiner, L.R.: A tutorial on hidden Markov models and selected applications in speech recognition. Proc. IEEE **77**, 257–286 (1989)
8. Chen, M.-Y., Kundu, A., et al.: Off-line handwritten word recognition using a hidden Markov model type stochastic network. IEEE Trans. Pattern Anal. Mach. Intell. **16**(5), 481–496 (1994). ISSN 0162–8828
9. Khadr, M.: Forecasting of meteorological drought using Hidden Markov Model (case study: the upper Blue Nile river basin, Ethiopia). Ain Shams Eng. J. **7**(1), 47–56 (2016). ISSN 2090–4479
10. Baum, L.E., Petrie, T., Soules, G., Weiss, N.: A maximization technique occurring in the statistical analysis of probabilistic functions of Markov chains. Ann. Math. Stat. **41**(1), 164–171 (1970)
11. Bilmes J.A.: A gentle tutorial of the EM algorithm and its application to parameter estimation for Gaussian mixture and hidden Markov models. U.C. Berkeley, TR-97–021 (1998). http://citeseer.ist.psu.edu/1570.html
12. Li, X., Parizeau, M., Plamondon, R.: Training hidden Markov models with multiple observations-a combinatorial method. IEEE Trans. Pattern Anal. Mach. Intell. **22**(4), 371–377 (2000)
13. Ye, F., Yi, N., Wang, Y.: EM algorithm for training high-order hidden Markov model with multiple observation sequences. J. Inf. Comput. Sci. **8**(10), 1761–1777 (2011)
14. Hadar, U., Messer, H.: High-order hidden Markov models? Estimation and implementation. In: Proceedings of the IEEE/SP 15th Workshop on Statistical Signal Processing, pp. 249–252 (2009)
15. Srivastava, A., Kundu, A., Sural, S., Majumdar, A.K.: Credit card fraud detection using hidden Markov model. IEEE Trans. Dependable Secure Comput. **5**(1), 37–48 (2008)
16. UCI machine learning repository. http://archive.ics.uci.edu/ml
17. https://www.lendingclub.com/info/download-data.action
18. Óskarsdóttir, M., et al.: The value of big data for credit scoring: enhancing financial inclusion using mobile phone data and social network analytics. Appl. Soft Comput. J. **74**, 26–39 (2018)
19. Hansen, P., Mladenovic, N., Perez, J.A.M.: Variables neighborhood search: methods and applications. Ann. Oper. Res. **175**, 367–407 (2010)
20. Mladenovic, N.: A variable neighborhood algorithm a new metaheuristics for combinatorial optimization. Abstracts of Papers Presented at Optimization Days, Montral, p. 112 (1995)

21. Hansen, P., Mladenovic, N.: Variable neighborhood search: principles and applications. Eur. J. Oper. Res. **130**, 449–467 (2001)
22. Hansen, P., Mladenovic, N.: Variable neighborhood search. In: Pardalos, P., Resende, M. (eds.) Handbook of Applied Optimization, pp. 221–234. Oxford University Press, London (2002)
23. Hansen, P., Mladenovic, N.: Tutorial on variable neighborhood search. Technical report G-2003-46, Les Cahiers du GERAD (2003)
24. Brimberg, J., Hansen, P., Mladenovic, N.: Convergence of variable neighborhood search. Technical report G-2003-45, Les Cahiers du GERAD (2003)
25. Garca, C.G., Prez, D., Garca, F.C.: Parallel variable neighborhood search for the linear ordering problem. In: Hansen, P., Mladenovic, N., Prez, J.A.M., Batista, B.M., Moreno-Vega, J.M. (eds.) Proceedings of the 18th Mini Euro Conference on Variable Neighborhood (2005)
26. Aupetit, S., Monmarch, N., Slimane, M.: Hidden Markov models training using population-based metaheuristics. In: Siarry, P., Michalewicz, Z. (eds.) Advances in Metaheuristics for Hard Optimization. Natural Computing Series, pp. 415–438. Springer, Heidelberg (2007). https://doi.org/10.1007/978-3-540-72960-0_20
27. Paul, D.B.: Training of HMM recognizers by simulated annealing. In: Proceedings of IEEE International Conference on Acoustics, Speech and Signal Processing, pp. 13–16 (1985)
28. Hamam, Y., Al Ani, T.: Simulated annealing approach for hidden Markov models. In: 4th WG-7.6 Working Conference on Optimization-Based Computer-Aided Modeling and Design, ESIEE, France (1996)
29. Chen, T.-Y., Mei, X.-D., Pan, J.-S., Sun, S.-H.: Optimization of HMM by the tabu search algorithm. J. Inf. Sci. Eng. **20**(5), 949–957 (2004)
30. Thomsen, R.: Evolving the topology of hidden Markov models using evolutionary algorithms. In: Guervós, J.J.M., Adamidis, P., Beyer, H.-G., Schwefel, H.-P., Fernández-Villacañas, J.-L. (eds.) PPSN 2002. LNCS, vol. 2439, pp. 861–870. Springer, Heidelberg (2002). https://doi.org/10.1007/3-540-45712-7_83
31. Rasmussen, T.K., Krink, T.: Improved hidden Markov model training for multiple sequence alignment by a particle swarm optimization - evolutionary algorithm hybrid. BioSystems **72**, 5–17 (2003)

A Multi-objective Metaheuristic for a Green UAV Grid Routing Problem

Elias L. Marques Jr.[1](✉), Vitor N. Coelho[1](✉), Igor M. Coelho[2](✉),
Bruno N. Coelho[3](✉), and Luiz S. Ochi[1](✉)

[1] Institute of Computer Science, Universidade Federal Fluminense, Niterói, Brazil
eliaslawrence.jr@gmail.com, vncoelho@gmail.com, satoru@ic.uff.br
[2] Computer Science Department, Universidade do Estado do Rio de Janeiro,
Rio de Janeiro, Brazil
igor.machado@ime.uerj.br
[3] REDEMAT – Rede Temática em Engenharia de Materiais,
Universidade Federal de São João del Rey, São João del Rei, Brazil
brunonazario@gmail.com

Abstract. This paper deals with Unmanned Aerial Vehicle (UAV) routing in dynamic grid scenarios with limited battery autonomy and multiple charging stations. The problem is inspired by real-world constraints,
specially designed for overcoming challenges of a limited vehicle driving
range. Recently, these kinds of vehicles have started to be used for deliver-
ing and collecting products, requiring experts in several knowledge fields
to manage this novel logistics. Inspired by a multi-criteria view of real
systems, we consider different objective functions introduced in the literature. A multi-objective variant of Variable Neighborhood Search is considered for finding a set of non-dominated solutions, while respecting the
navigation over forbidden areas and also battery capacity. A case of study
was developed where one UAV has to attend clients spread throughout
a grid representing a map. The drone starts in a given grid point with a
given battery charge, where the grid is composed by four different kinds
of points: a regular one and three special (prohibited, recharge and client
delivery). Any sequence of valid adjacent points forms a route, but since
this yields a huge number of combinations, a pre-processing technique is
proposed to pre-compute distances in a given dynamic scenario. Computational results demonstrate the performance of different variants of the
proposed algorithm.

Keywords: Unmanned Aerial Vehicle · Microgrids · Multi-objective
optimization · Variable Neighborhood Search

1 Introduction

Technological innovations such as the miniaturization of electronic control systems and the cost reduction of electronic components [8] has resulted in an

© Springer Nature Switzerland AG 2020
R. Benmansour et al. (Eds.): ICVNS 2019, LNCS 12010, pp. 152–166, 2020.
https://doi.org/10.1007/978-3-030-44932-2_11

upsurge in the availability of Unmanned Aerial Vehicles (UAV), also known as Drones, or Unmanned Aerial Systems (UAS).

Although the drone is often related to hobbyists, entertainment and photographic industries, their uses has been spread to military, civil and commercial applications. Aerial surveillance, recognizing and tracking objects are some of many other applications that are emerging with potential for UAV's use. Countless others may arise from human creativity in the near future [3]. There are already some works that show applications in everyday life:

- Infrastructure inspection [13,15,16]: The UAV can follow a predetermined path or could move by visual servoing and detects by the means of image treatment the size and location of defects and cracks. The use of a drone for inspection will give us multiple advantages over traditional inspection method:
  - Reducing work accident risk;
  - Budget reduction: less logistics and less working hours;
  - Less invasive operations: the bridge will not be closed for traffic while the inspection is done.
- Power Line Inspection [1,5]:
  - Aerial inspection of electric power transmission lines is typically performed using human-piloted helicopters, which is a procedure that is both expensive and prone to accidents, bringing risks to human beings' lives. In this sense, the drone is a low cost solution with several potential benefits.
- Surveillance of a target space using aerial vehicles is a topic of current research interest for applications such as weather monitoring, geographical surveys, and perhaps extraterrestrial exploration [18];
- The great flexibility of UAVs can enable new approaches during collection of remote sensing data, which for example integrate real-time mapping and autonomous navigation [11];
- The environmental monitoring is the wide research field for single UAV solutions, where the monitoring of the environment is realized just by one vehicle [12].

The load transportation sector, in particular, is already showing some interest and investments in UAV applications. The growth of e-commerce has sustained this interest from huge companies. Transportation drones are capable of safely taking off and landing in the proximity of buildings and humans, improving the quality of current service in congested or remote areas [8].

When deliver service are discussed, we, implicitly, are talking about a Traveling Salesman Problem (TSP) and its variations, as the Vehicle Routing Problem for example. Which briefly means a problem of designing optimal routes from one or several depots to a number of geographically scattered cities, customers or strategic points, subject to side constraints [14].

Although there are many works in the literature related to TSP variations [10], the ones that approach UAV routing are still few as the TSPD (Traveling Salesman Problem with Drone) [2], the vehicle routing problem with drones [23] and the VNS approach of Schermer et al. [20].

However, we can not just focus on technological advancements and simply forget about the damage to the environment that they may cause. That is why the scientific community has been concerned so much with developing green technologies and this does not differ in computing. The Green Vehicle Routing Problem (G-VRP), for example, proposed by Erdoğan et al. [7], add to the original VRP, constraints about fuel economy.

The TSP with hotel selection [22] is a TSP variation with similarities with the problem addressed in this article. The main objective is to minimize the number of trips and total traveled time. This problem is found in real scenarios like delivery of products by electric vehicles that need to be recharged along a tour.

In order to approach real systems, it is necessary to design a multi-objective problem providing a set of non-dominated solutions with different possible routes and schedules. The more objective functions and constraints, the more similar to real world, but also more complex the problem becomes.

The main contributions of this current work are:

– Complement to the linear mathematical model of Coelho et al. [3] by developing a metaheuristic algorithm for a time-dependent UAV routing problem, in particular:
  • respecting UAVs operational requirements;
  • tackling the micro-airspace considering a scenario of points inspection, and avoiding prohibited points (docking constraints) [4];
  • integrating UAVs into the new concepts of mini/microgrid systems, in which vehicles can be charged at different points of the future smart cities;
  • dynamic routes considering drones already in movement: instances with initial battery different than 100%, random origin point and a number of clients already visited.

The remainder of this paper is organized as follows. Section 2 describe the proposed model and the range of real parameters considered in our analyses, while Sect. 3 contains the methodology employed to solve the problem. In Sect. 4, one can find the computational experiments comparing the different implementations, instances, variables and results. Finally, Sect. 5 concludes the work and presents future research directions.

2 Problem Description

The case of study designed here is composed of an airspace divided into horizontal and vertical strips, where the vehicle is allowed to move following Chebyshev distance, where the distances between any adjacent points are the same. Energy stations are spread in the routing area and accessed by the drone for recharging its batteries. To represent prohibited areas, the grid is also composed by prohibited points which the UAV cannot access, otherwise it would invalidate the route.

As a routing problem, the vehicle should attend clients that are spread across the grid. To do so, the point correspondent to the client should be part of the final route. That means that the coordinates X and Y of the client must be part of the array that represents the solution. On Sect. 3 we discuss an alternative to preprocess shortest distances and store them in auxiliary data structure, so that only the origin-destination points and distances need to be considered between charging and delivery points. However, due to dynamic nature of the problem, it may be required to process these once again, after changes occur in the input data[1].

2.1 Objectives

As previously mentioned, this paper addresses a multi-objective problem, in which we want to find a set of non-dominated solutions. In a real UAV routing problem, a great amount of variables should be considered in order to obtain the most adequate solution. However, in order to turn this real problem into a computable one, we mapped three main objectives that summarize, in a satisfactory manner, how the real system should behave.

Consumption: it is desirable that during the route, the vehicle consume the least possible of battery/fuel.

Final charge: it is interesting to finish the route with the maximum charge rate possible, ensuring that the drone is prepared for a future route.

Time: the total route should be performed in the shortest possible time frame.

As can be seen, the algorithm proposed in this paper focuses on finding a balanced trade-off between solutions, since one element may affect another one. Smaller the time, greater the velocity, than, greater the consumption. Greater the final charge means more time spent recharging/fueling which results in greater times.

2.2 Constraints

In order to have a valid solution, the route must attend some requirements listed below:

Consumption: The fuel/battery level of the vehicle should not reach below zero in any part of the path, that would mean that the UAV would be out of fuel/energy in the middle of the route. However, if it reaches zero and the route is over or the drone reached one energy point, this does not affect the validity of the solution.

Prohibited Area: In real life, there are areas where drones are not allowed to access or cross. This situation was represented by special points scattered across the grid. If the route contains these points, the solution is invalid.

---

[1] The current work considers that the dynamic data is passed as input, so that no changes need to be performed during the search. As instances already consider arbitrary drone initial location and capacity (battery load), a time-dependent variant can be considered as an extension of this work (see Sect. 5).

2.3   Variables

There are two variables that affect the final result of the objective functions: speed in the stretch and charging time.

Speed in the stretch: influence not just the total time of the route, but also the consumption, once higher the speed (v), greater the consumption. The fuel/battery level at the end of the stretch (f) is a result of the fuel/battery level at the beginning of the stretch (f_0) decreased by the fixed consumption (c_f) and the speed multiplied by the coefficient of variable consumption (c_v) as shown at Eq. 1.

$$f = f_0 - v \times c_v - c_f \tag{1}$$

Time at energy station: the time spent at the energy station (r) is added to the total time spent at the route. However, if the vehicle spends more time at it, it can accumulate more fuel/energy to its battery. The fuel/battery level at the end of the stretch is a result of the fuel/battery level at the beginning of the stretch increased by the quantity of fuel/energy recharged (f_r) as shown at Eq. 2.

$$f = f_0 + f_r \tag{2}$$

Equation 3 shows that the time at the end of the stretch (t) is a result of the time at the beginning of the stretch (t_0) increased by the quantity of fuel/energy recharged multiplied by the coefficient of time per fuel/energy (t_f) as shown at Eq. 3.

$$t = t_0 + f_r \times t_f \tag{3}$$

3   Methodology

The metaheuristic GRASP with MOVND (G-MOVND) as local search was the chosen to be applied. It can be seen as a multi-start metaheuristic for combinatorial optimization problems, in which each iteration basically consists of two phases: construction and local search.

3.1   Construction

The construction phase is based on GRASP, where "[...] construction phase builds a solution using a greedy randomized adaptive algorithm" [19]. In a greedy construction of this problem we iteratively choose the closest client from the actual position. However, the construction phase of this algorithm is a greedy procedure with random and adaptative components, meaning that instead of always choosing the closest one, we select a set of k clients closest to the actual position and, then, we choose a client randomly from this set (candidate list) to be inserted in the initial solution. In this work, we get to the parameter $k = 3$ in a empirical way.

3.2   VND Local Search

The Variable Neighborhood Descent (VND), developed by Mladenović and Hansen [17] in 1997, is a method and a method that has proven to be very efficient and capable of navigating the space of solution based on the use of neighboring solutions. Local search heuristics commonly use one neighborhood structure (for example, Hill Climbing), unlike VND. This metaheuristic works by applying a local search method concerning a neighborhood structure to a initial solution x. If the solution x' obtained is better than the former, we attribute x' to x (x := x'), and continue the search with the current neighborhood structure; otherwise, we change it.

The multi-objective variant of the VND (MOVND) [6] was implemented with 10 neighborhood structures. Different neighborhoods affect different objective functions. As we were dealing with a set of solutions and not just a single desired final solution, we do not use just one best that is modified through the iterations, but a pool of solutions. An initial solution is generated by the GRASP, which now passes through exhaustive VND searchers, the obtained neighboring solutions are inserted into a global pool if they are not dominated.

The neighborhoods are presented next, in the order of execution of the metaheuristic:

Swap. This neighborhood is responsible to switch clients positions in the route. If the solution generated is not dominated or equal to any other route in the current pool of solutions, the new solution is added to it.

As this is the more costly neighborhood, were implemented other versions, so it was possible to compare the one that provides best results during the computational experiments. The difference from the versions was that instead of switching a client from the route with every other, the switch just occur with the k nearest, furthest or random clients.

Remove Recharge. This neighborhood is responsible for remove possible recharge points in each subpath (path between clients) and verify if it improves the current solution. The idea is that removing this stretch, the UAV would reach the end of the route in a shorter time.

We find a path that links two clients or the origin to a client and remove it from the route. After, we link directly the two clients. This way if there was a recharge point in this subpath it would be removed.

Closest Recharge. After we remove unnecessary recharge points, we try to add other ones that could improve the current solution.

Remove Repeated. After all the previous operations been executed on the route, we might have inserted repeated clients in the route. The idea of this neighborhood is to remove the repeated ones, trying to reduce the size of the route and, in consequence, reduce the consumption and/or time.

Speed Section Increase. Increases speed by 1 unit in each entire section of the route that links two important points (clients and/or recharge point).

Speed Section Decrease. Decreases speed by 1 unit in each entire section of the route that links two important points (clients and/or recharge point).

Speed Random Increase. Increases speed by 1 unit in each segment of a subpath chosen randomly.

Speed Random Decrease. Decreases speed by 1 unit in each segment of a subpath chosen randomly.

Recharge Random Increase. Increases by 1 unit the load percentage at each recharge point of a subpath chosen randomly.

Recharge Random Decrease. Decreases by 1 unit the load percentage at each recharge point of a subpath chosen randomly.

3.3   Acceptance Criterion

After generating a neighbor of the current solution, the current route is evaluated and compared to the routes in the pool of solutions. It will be inserted if it is not dominated by any other one present in the pool. If the new route dominates any other, the latter is removed from the set of solutions.

In this case of study, we limited the size of the set of non-dominated solution, limiting, in consequence, the number of operations (local search). If the set is full and the new route is non-dominated, it will only be inserted if there is a solution in the pool with result smaller than the new one. Thus, the incoming solution should dominate, at least, one solution of the current pool.

3.4   VND Implementation

To solve the proposed problem by generating the routes, different versions of an algorithm were implemented in C++. Four versions of the SWAP method were implemented, since it is the most computationally costly neighborhood. The versions, as described in a section before, were named S1, S2, S3 and S4. The size of the pool was also considered as a variable and different tests were done to verify its impact to the results. It was considered pools of size 5, 15 and 30.

The MOVND was the metaheuristic chosen, as addressed in previous sessions. However, a different approach was also considered. The original MOVND with a pool of solutions is known for execute a loop for each solution and inside a loop for each neighborhood. In our current implementation, there is an inversion where the external loop runs through the neighborhoods and the internal one runs through the solutions. This metaheuristic was named I-MOVND.

Algorithm 1. I-MOVND

1: **procedure** I-MOVND($E, Neighborhood$)
2:    **repeat**
3:       $x \leftarrow Select(E \setminus Si)$ ▷ Selection among the non-exploited points
4:       **for all** $k \in Neighborhood$ **do**
5:          $added \leftarrow true$
6:          **while** $added$ **do**
7:             $added \leftarrow false$
8:             $x' \leftarrow neighbor(x, k)$
9:             **if** MO-Improvement(E, x, x') **then**
10:                $added \leftarrow true$
11:                $E \leftarrow Update(E, x')$
12:             **end if**
13:          **end while**
14:       **end for**
15:    **until** $E \setminus Si = \emptyset$
16:    **return** E
17: **end procedure**

The activity of calculate routes can be really costly when executed many times through the algorithm. Therefore, a pre-processing was implemented in order to make comparisons if it brings gains to the results. The method consists of pre-calculating the best routes between the important points of the map (clients, prohibited area and recharge points). The routes are saved and after read as part of the entry of the problem.

To summarize the different implementations, each sample varies as:

– Instance
   • eil51A and eil51B: both with 51 clients, 5% of prohibited points and 1% of recharge points. The difference between each other is due to the position of the random prohibited and recharge points generated.
   • eil101A and eil101B: same as before but with 101 clients
– SWAP
   • S1: swap 1
   • S2: swap 2
   • S2-3: swap 2 and 3
   • S2-4: swap 2 and 4
   • S4: swap 4
– Local Search
   • MOVND
   • I-MOVND
– Pre-processing
   • Yes
   • No

4 Computational Experiments

As, from our knowledge, there is not a well-established library of instances for the problem addressed in this article, with all the restrictions considered here in the literature, two well known TSP instances in Euclidean 2D format were used as base of the tests: Christofides/Eilon eil51 and eil101. These instances have 51 and 101 clients, respectively. From these, arrays were generated representing the area that comprises all clients, where the value of x varies from the minimum x value from the instance to the maximum. The same happens to the y coordinates.

After the matrix is generated, points in it are chosen randomly and set as points of recharge and prohibited ones. For matter of tests, it was stipulated a rate of 5% of the matrix to prohibited points and 1% to recharge points. For each instance, two different maps were generated.

Every sample was executed 5 times, each one for 5 min. It was considered the set of non-dominated solutions of all the 5 executions in matter of results.

4.1 Comparison Measures

In order to compare the results obtained during the experiments, three measures referents to the quality of the solutions were used: hypervolume, coverage and cardinality.

According to [21], hypervolume is an indicator associated with an approximation given by the volume of the objective space portion that is weakly dominated by a set. This indicator needs the specification of a reference point Z that denotes an upper bound over all the objectives. In this problem normalized objective function were used. In addition, it has been ensured that as closer to better the measure is. The code of the calculation of hypervolume is provided by [9].

The other measures are coverage and cardinality. Coverage, meaning the percentage amount of solutions generated by a specific method that is in the Pareto reference. For example, if we would like to compare methods A and B, we would run both methods and in the end we would gather the solutions of both methods and select the only ones not dominated. In this final pool, we have 4 solutions from method A and 6 solutions of method B. Then, the coverage of A is of 40% and 60% of B.

Cardinality is referent to the absolute amount.

4.2 Computational Results

The first tests were developed to verify the methods that return the best results. To do so, it was used samples with pre-processing and pool of fixed size of 30.

The best results ranged between the S2-4 and S4 from VND and I-VND as shown at Table 1. Therefore, the next step was to execute the algorithm in order to evaluate how the size of the pool and the pre-processing affected the result.

Table 1. Samples with pre-processing.

Methods	eil51P5R1a			eil101P5R1a		
	Hyperv.	Cov.	Card.	Hyperv.	Cov.	Card.
MOVND S1	0.37	0.02	1	0.26	0	0
MOVND S2	0.10	0	0	0.29	0	0
MOVND S2-S3	0.27	0	0	0.30	0	0
MOVND S2-S4	0.40	0	0	**0.51**	**0.45**	**45**
MOVND S4	**0.46**	**0.60**	**29**	0.47	0	0
I-MOVND S1	0.40	0	0	0.20	0	0
I-MOVND S2	0.26	0	0	0.31	0	0
I-MOVND S2-S3	0.33	0	0	0.37	0.22	22
I-MOVND S2-S4	0.43	0.35	17	0.46	0	0
I-MOVND S4	0.42	0.02	1	0.48	0.34	34

Table 2. Sample eil51P5R1a.

Methods	With PP			Without PP		
	Hyperv.	Cov.	Card.	Hyperv.	Cov.	Card.
MOVND 30 S2-S4	0.30	0	0	0.30	**0.23**	**22**
MOVND 30 S4	**0.35**	**0.26**	**24**	0.29	0.01	1
I-MOVND 30 S2-S4	0.32	0.18	17	0.29	0	0
I-MOVND 30 S4	0.31	0.01	1	0.29	0	0
MOVND 15 S2-S4	0.26	0	0	0.27	0	0
MOVND 15 S4	0.32	0.15	14	0.28	0	0
I-MOVND 15 S2-S4	0.30	0.03	3	0.28	0	0
I-MOVND 15 S4	0.27	0	0	**0.31**	0.12	11
MOVND 5 S2-S4	0.23	0	0	0.27	0	0
MOVND 5 S4	0.21	0	0	0.29	0	0
I-MOVND 5 S2-S4	0.30	0	0	0.26	0.01	1
I-MOVND 5 S4	0.24	0	0	0.27	0	0

The Tables 2, 3, 4, 5 show that pre-processing results in better values of hypervolume, coverage and cardinality in almost all the tests. Meaning that the pre-calculation of routes generates a pool of solutions more diversified which is a good result when we are talking about a multiobjective problem. In terms of max size of the pool, in the samples with the smaller instance, bigger the pool, better the results. And with the bigger instance, this difference was not so evident since the size of the instance already orders more computational time and a bigger pool does the same.

Table 3. Sample eil51P5R1b.

Methods	With PP			Without PP		
	Hyperv.	Cov.	Card.	Hyperv.	Cov.	Card.
MOVND 30 S2-S4	0.23	0.29	24	**0.21**	0	0
MOVND 30 S4	0.21	**0.33**	**27**	0.19	0	0
I-MOVND 30 S2-S4	0.19	0.13	11	0.20	0	0
I-MOVND 30 S4	**0.35**	0.06	4	0.16	0	0
MOVND 15 S2-S4	0.17	0	0	0.17	0	0
MOVND 15 S4	0.18	0	0	0.16	0	0
I-MOVND 15 S2-S4	0.22	0.16	13	0.15	0	0
I-MOVND 15 S4	0.19	0	0	0.15	0	0
MOVND 5 S2-S4	0.14	0	0	0.12	0	0
MOVND 5 S4	0.10	0	0	0.15	0	0
I-MOVND 5 S2-S4	0.15	0	0	0.16	0	0
I-MOVND 5 S4	0.16	0	0	0.14	0	0

Table 4. Sample eil101P5R1a.

Methods	With PP			Without PP		
	Hyperv.	Cov.	Card.	Hyperv.	Cov.	Card.
MOVND 30 S2-S4	**0.43**	0.30	29	0.14	0	0
MOVND 30 S4	0.40	0	0	0.03	0	0
I-MOVND 30 S2-S4	0.40	0	0	0.06	0	0
I-MOVND 30 S4	0.41	0.21	20	0.09	0	0
MOVND 15 S2-S4	0.29	0	0	0.11	0	0
MOVND 15 S4	0.41	0.125	12	0.12	0	0
I-MOVND 15 S2-S4	0.43	**0.36**	**35**	0	0	0
I-MOVND 15 S4	0.36	0	0	0.10	0	0
MOVND 5 S2-S4	0.33	0	0	0.15	0	0
MOVND 5 S4	0.35	0	0	0.27	0	0
I-MOVND 5 S2-S4	0.38	0	0	**0.32**	0	0
I-MOVND 5 S4	0.35	0	0	0.23	0	0

On Figs. 1 and 2 we can see the pareto fronts generated by the execution of the algorithm. We can infer from it that the pool of size 30 generates better results to bigger instances, while the pool of maximum size 15, generates better results with smaller instances (less points of inspection/clients), smaller pools.

Table 5. Sample eil101P5R1b.

Methods	With PP			Without PP		
	Hyperv.	Cov.	Card.	Hyperv.	Cov.	Card.
MOVND 30 S2-S4	0.17	0	0	0	0	0
MOVND 30 S4	0.08	0	0	0.04	0	0
I-MOVND 30 S2-S4	0.09	0	0	0.03	0	0
I-MOVND 30 S4	0.17	0	0	0.04	0	0
MOVND 15 S2-S4	**0.24**	**0.79**	**15**	0.14	0	0
MOVND 15 S4	0.18	0	0	0.02	0	0
I-MOVND 15 S2-S4	0.22	0	0	0.04	0	0
I-MOVND 15 S4	0.19	0	0	0.01	0	0
MOVND 5 S2-S4	0.12	0	0	0.09	0	0
MOVND 5 S4	0.12	0	0	0.16	0	0
I-MOVND 5 S2-S4	0.16	0	0	0.11	0	0
I-MOVND 5 S4	0.11	0	0	**0.19**	0	0

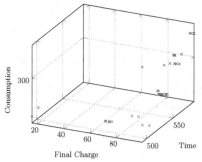

(a) Instance eil101 70% of battery 50% of clients visited.

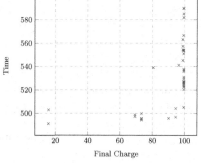

(b) Final Charge x Time: VND-15 (blue) and VND-30 (red)

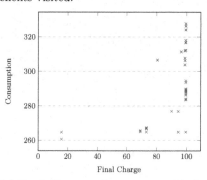

(c) Final Charge x Consumption: VND-15 (blue) and VND-30 (red)

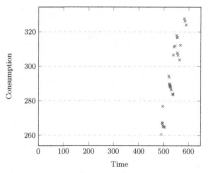

(d) Time x Consumption: VND-15 (blue) and VND-30 (red)

Fig. 1. Pareto fronts for instance eil101-70-50

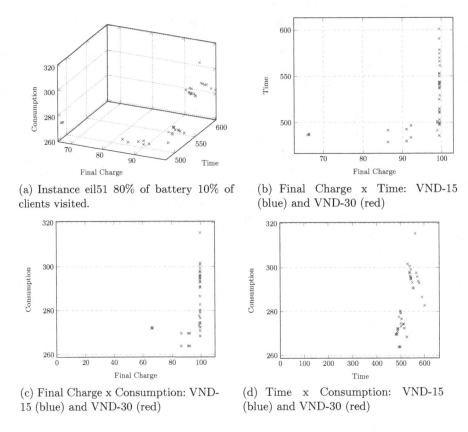

(a) Instance eil51 80% of battery 10% of clients visited.

(b) Final Charge x Time: VND-15 (blue) and VND-30 (red)

(c) Final Charge x Consumption: VND-15 (blue) and VND-30 (red)

(d) Time x Consumption: VND-15 (blue) and VND-30 (red)

Fig. 2. Pareto fronts for instance eil51-80-10

5   Conclusions

The multi-objective, grid, docking constraint, and the concern with consumption (Green Computing) and the dynamism of this problem shows a very practical approach, for real applications. Instances have considered arbitrary initial positions of drones, and also variable initial battery capacities, so that it is possible to integrate this tool into an online solver that resolves a series of instances considering changes on the dynamic features of the scenario (due to wind conditions, logistics and other operational constraints).

This work also shows a potential of growth approaching increasingly from a real situation. Thus, we visualize future works taking into account more layers of grid representing the vertical movement of the drone. Besides that a problem with heterogeneous drones and temporary prohibited points are some other variables that could be explored in future works involving this problem.

Acknowledgment. Vitor N. Coelho would like to thank the Brazilian agency FAPERJ (E-26/202.868/2016). Luiz S. Ochi was supported by FAPERJ and CNPq

(301593/2013-2), Igor M. Coelho and Elias L. Marques Jr. by FAPERJ. This study was financed in part by the Coordenação de Aperfeiçoamento de Pessoal de Nível Superior – Brasil (CAPES) - Finance Code 001.

References

1. Adabo, G.J.: Long range unmanned aircraft system for power line inspection of Brazilian electrical system. J. Energy Power Eng. **8**(2), 394–398 (2014)
2. Agatz, N., Bouman, P., Schmidt, M.: Optimization approaches for the traveling salesman problem with drone. Transp. Sci. **52**(4), 965–981 (2018)
3. Coelho, B.N., et al.: A multi-objective green UAV routing problem. Comput. Oper. Res. **88**, 306–315 (2017)
4. Coelho, V.N., Grasas, A., Ramalhinho, H., Coelho, I.M., Souza, M.J., Cruz, R.C.: An ILS-based algorithm to solve a large-scale real heterogeneous fleet VRP with multi-trips and docking constraints. Eur. J. Oper. Res. **250**(2), 367–376 (2016)
5. Deng, C., Wang, S., Huang, Z., Tan, Z., Liu, J.: Unmanned aerial vehicles for power line inspection: a cooperative way in platforms and communications. J. Commun. **9**(9), 687–692 (2014)
6. Duarte, A., Pantrigo, J.J., Pardo, E.G., Mladenovic, N.: Multi-objective variable neighborhood search: an application to combinatorial optimization problems. J. Glob. Optim. **63**(3), 515–536 (2014). https://doi.org/10.1007/s10898-014-0213-z
7. Erdoğan, S., Miller-Hooks, E.: A green vehicle routing problem. Transp. Res. Part E: Logist. Transp. Rev. **48**(1), 100–114 (2012)
8. Floreano, D., Wood, R.J.: Science, technology and the future of small autonomous drones. Nature **521**(7553), 460 (2015)
9. Fonseca, C.M., Paquete, L., López-Ibánez, M.: An improved dimension-sweep algorithm for the hypervolume indicator. In: 2006 IEEE International Conference on Evolutionary Computation, pp. 1157–1163. IEEE (2006)
10. Gutin, G., Punnen, A.P.: The Traveling Salesman Problem and Its Variations, vol. 12. Springer, Boston (2006). https://doi.org/10.1007/b101971
11. Haala, N., Cramer, M., Weimer, F., Trittler, M.: Performance test on UAV-based photogrammetric data collection. Proc. Int. Arch. Photogram. Remote Sens. Spat. Inf. Sci. **38**(1/C22), 7–12 (2011)
12. Harris, A., Sluss, J.J., Refai, H.H., LoPresti, P.G.: Alignment and tracking of a free-space optical communications link to a UAV. In: The 24th Digital Avionics Systems Conference, DASC 2005, vol. 1, pp. 1–C. IEEE (2005)
13. Irizarry, J., Gheisari, M., Walker, B.N.: Usability assessment of drone technology as safety inspection tools. J. Inf. Technol. Constr. (ITcon) **17**(12), 194–212 (2012)
14. Laporte, G.: The vehicle routing problem: an overview of exact and approximate algorithms. Eur. J. Oper. Res. **59**(3), 345–358 (1992)
15. Máthé, K., Buşoniu, L.: Vision and control for UAVs: a survey of general methods and of inexpensive platforms for infrastructure inspection. Sensors **15**(7), 14887–14916 (2015)
16. Metni, N., Hamel, T.: A UAV for bridge inspection: visual servoing control law with orientation limits. Autom. Constr. **17**(1), 3–10 (2007)
17. Mladenović, N., Hansen, P.: Variable neighborhood search. Comput. Oper. Res. **24**(11), 1097–1100 (1997)
18. Nigam, N., Kroo, I.: Persistent surveillance using multiple unmanned air vehicles. In: 2008 IEEE Aerospace Conference, pp. 1–14. IEEE (2008)

19. Resende, M.G.C., Ribeiro, C.C.: Greedy randomized adaptive search procedures: advances and extensions. In: Gendreau, M., Potvin, J.-Y. (eds.) Handbook of Metaheuristics. ISORMS, vol. 272, pp. 169–220. Springer, Cham (2019). https://doi.org/10.1007/978-3-319-91086-4_6
20. Schermer, D., Moeini, M., Wendt, O.: A variable neighborhood search algorithm for solving the vehicle routing problem with drones. Technical report, Technische Universität Kaiserslautern (2018)
21. Talbi, E.G.: Metaheuristics: from Design to Implementation, vol. 74. Wiley, Hoboken (2009)
22. Vansteenwegen, P., Souffriau, W., Sörensen, K.: The travelling salesperson problem with hotel selection. J. Oper. Res. Soc. $63(2)$, 207–217 (2012)
23. Wang, X., Poikonen, S., Golden, B.: The vehicle routing problem with drones: several worst-case results. Optim. Lett. $11(4)$, 679–697 (2016). https://doi.org/10.1007/s11590-016-1035-3

A VNS Approach for Batch Sequencing and Route Planning in Manual Picking System with Time Windows

Jerzy Duda$^{(\boxtimes)}$ and Adam Stawowy

Department of Applied Computer Science, Faculty of Management, AGH University of Science and Technology, Gramatyka St. 10, 30-067 Krakow, Poland
{jduda,astawowy}@zarz.agh.edu.pl

Abstract. In this paper an order picking and route planning problem is studied. The main objective is to minimize the number of pickers involved in collecting the order pool and the total distance covered by the pickers. Since the problem under study is NP-hard, a variable neighborhood search (VNS) is proposed as a heuristic solution approach. Neighborhood is changed according to VNS scheme employing four tailored structures.

Finally, computational tests demonstrate that the proposed VNS algorithm can find good quality solutions for all practical problems examined. The objective values, regarding both the number of pickers employed and the total distance covered by them, are better than the results of genetic algorithm and close to the ones obtained by CPLEX Solver, if it was able to provide a feasible solution.

Keywords: Order picking · Batch sequencing · Variable neighborhood search

1 Introduction

Warehouses are essential elements in supply chains and logistics as they are responsible for linking suppliers, production facilities and distribution systems, as well as for storing products and handling customer orders on time. Since warehousing can account for approximately 20% of supply chain costs [8], any improvement in the efficiency of warehouse operations can generate significant savings in operational costs.

The problem of order-picking is defined as collecting the items from their location in the warehouse in order to satisfy the demand from internal or external customers [12]. Petersen [12] also distinguishes two types of the order-picking problem in which human resources (pickers) are involved:

- items-to-picker, in which the items are automatically delivered to the picker at the point of collection and dispatching of items,
- picker-to-items, in which an employee (picker) travels to different locations in order to pick up the demanded items.

© Springer Nature Switzerland AG 2020
R. Benmansour et al. (Eds.): ICVNS 2019, LNCS 12010, pp. 167–177, 2020.
https://doi.org/10.1007/978-3-030-44932-2_12

In many warehouses we deal with the second type of the order-packing problem. Despite the increasing automation of the process, according to the literature, still around 80% warehouses in Western Europe use manually picker-to-items system [14].

Order picking consists of three successive stages: (1) order batching, (2) batch sequencing, and (3) picker routing. Order batching groups many customer orders into a single picking order taking into account, for example, the elements common to customer orders, delivery times of client orders, and the location of items in the warehouse. Order batching is especially important in manual picking systems, because grouping several customer orders into individual picking orders reduces the total number of routes, total travel distance or total travel time. Batch sequencing determines the order in which batches are processed and how the clients' orders are allocated to operators (pickers) to meet deadlines and minimize tardiness and earliness of customer orders. Picker routing consists in planning the best route for all pickers to follow in order to retrieve all the elements of batch, starting and ending in the depot.

The warehouses may be organized in different ways. We consider a standard layout where the bays (storage locations) of identical size are arranged on both sides of picking aisles. Order pickers can move from one aisle to another by two cross aisles, one at the front and one at the rear of the picking area. A wide-aisle warehouse is assumed, i.e. the width of picking aisles allows for overtaking maneuvers, so traffic jams caused by the pickers collecting the items from the same storage location are not considered. However, in the considered case, the depot is not a single point but a line located at front aisle where the pickers return to in order to deposit the picked items.

In this paper we study the joint batch sequencing and picker routing problem with time windows, i.e. the orders should be completed during given period of time (release time of an order and due date required by a customer), at minimal costs. The aim of the paper is to develop the model and an appropriate algorithm of its solution able to achieve optimized pickers' tours. We propose a variable neighborhood search algorithm with four neighborhood structures to solve the problem. Test cases based on data from one of the automotive warehouses in Poland are used to evaluate the solutions from the proposed heuristic and are compared with the outputs from genetic algorithm and CPLEX Solver.

The paper has the following structure. Section 2 presents the literature review. Definition of the problem and notation are described in Sect. 3. Section 4 gives the details on proposed heuristic approach. The computational experiments are summarized in Sect. 5, and the conclusions are drawn in Sect. 6.

2 Literature Review

There are many studies reported in the literature related to one or more stages of the order picking process (subproblems), i.e. order batching, batch sequencing and picker routing. A comprehensive review on problems' formulations as well as solution approaches are presented in [3, 5] and [8]. The surveys' authors characterize models with various objective functions, consideration of due dates, warehouse layouts and information availability. They classify the solution approaches into five categories: (1) simulation based, (2) exact methods, (3) heuristic (greedy) algorithms, (4) metaheuristics, and (5) data mining methods. As each subproblem of the order picking process is NP-hard and realistically sized instances of them cannot be solved in polynomial time, the authors note the great popularity of metaheuristics.

For practical reasons, this section gives a limited review of the methods used to solve considered problem: we focus only on approaches that use VNS metaheuristic or any of its variants.

Albareda-Sambola et al. [1] addressed basic order batching problem with the objective of minimizing the total travel time. They proposed a heuristic based on Variable Neighborhood Search (VNS), where three neighborhoods were employed: (1) transferring an order from one batch to another one, (2) transferring at most two orders from one batch to others, and (3) transferring at most two orders. The results of simulation experiments showed that proposed approach is competitive method which can be used in practice. It should be noted that the test instances were limited in size (max 250 orders) and VNS algorithm was compared only with Clarke and Wright constructive heuristic.

Henn [7] proposed VND and VNS to minimize the total tardiness of customer orders for order batching and batch sequencing. In both approaches the author used a constructive Earliest Start Date (ESD) rule to generate an initial solution. Two classes of neighborhoods were applied: (1) four moves that change the position of a complete batch, and (2) four moves that change the position of a customer order. In both cases only moves generating feasible solutions were considered. Numerical experiments showed that both approaches generate better solutions than those created by heuristics based on priority rules. In addition, they showed that the deterministic VND algorithm outperforms the stochastic VNS. The VNS approach is able to generate high quality solutions if the calculation time is long enough. However, implementing these algorithms can significantly improve warehouse performance.

Menéndez et al. [9] suggested a General Variable Neighborhood Search (GVNS) algorithm to tackle the order batching problem which objective was to minimize the total time needed to collect all items. They used two neighborhood structures: (1) insertion of any order in a different batch, and (2) swap of two orders from different batches. The perturbation was achieved by random swapping two orders from different batches. Both neighbor and perturbation moves were performed in such a way that each result was a feasible solution. The authors confronted the operation of their approach with the Albareda-Sambola et al. algorithm [1] using data sets presented by the authors of the latter algorithm. The results indicated that GVNS algorithm outperforms VND approach of Albareda-Sambola et al.

More recently, Menéndez et al. [10] developed a parallel GVNS for the min–max order batching and sequencing problem which objective is to minimize the maximum retrieving time for any batch. They used two neighborhood structures, the same as were presented in [9]. The shake stage involved four different orders allocated in four different batches and was inspired by the ejection chain method. The best improvement local search, based on swap moves, was employed as a local search. The parallel computing was performed in a basic form: the algorithm used several threads, where each one was responsible for perturbation the current solution with the shake procedure. The objective of the parallelization was not to reduce the computing time, but to explore a wider area of the solution space. Additionally, the authors introduced a novel scheme for calculating the objective function (based on a matrix of retrieving times) which twice reduced the computing time. The extensive experimental comparison showed that the parallel version of the VNS evidently outperformed the best known at this time approaches.

Scholz et al. [13], for the first time in literature, considered all picking subproblems simultaneously. They introduced a mathematical model of the joint order batching, sequencing and routing problem that allows for exact solving small problem instances. For larger instances, a VND algorithm was presented. The initial solution for the VND was generated by means of two constructive approaches as the better from two solutions obtained. The first approach was ESD rule introduced by Henn [7] and the second – a seed algorithm which consecutively rearranged the orders to minimize the total tardiness. The authors' algorithm employed six neighborhoods: (1) exchanging two batches, (2) breaking up a complete batch and reassign the orders to other batches, (3) moving an order from one batch to another batch assigned to the same picker, (4) moving an order from one batch to another batch assigned to another picker, (5) exchanging two orders which are included in different batches assigned to the same picker, and (6) exchanging two orders which are included in different batches assigned to different pickers. By means of extensive numerical experiments, the authors demonstrated that the algorithm provided solutions of very good quality. Furthermore, it was shown that a simultaneous solution approach to the picking subproblems can reduce the total tardiness of all customer orders by up to 84% which makes the proposed VND a valuable tool for an efficient organization of warehouse operations.

Ardjmand et al. [2] modelled and solved the joint order batching, sequencing and routing problem with multiple pickers in a wave picking warehouse of a major US third party logistics company. For small size waves a Lagrangian decomposition heuristic combined with a particle swarm optimization (LD-PSO) algorithm was proposed, while for large-scale problems a hybrid parallel simulated annealing and an ant colony optimization (PSA-ACO) was presented. The approaches were compared against the heuristic being used in the examined warehouse and a VND algorithm proposed by Scholz et al. [13] and it was shown that PSA-ACO and VND can improve the makespan by approximately 7.0% over the existing heuristics.

Finally, Gil-Borrás et al. [5] considered the online order batching problem (OOBP) in which changes of pick-lists during a pick cycle are allowed. The objective was to minimize the maximum time that the orders remained in the system. To tackle this problem the authors proposed the Basic Variable Neighborhood Search (BVNS) approach which combines stochastic and deterministic exploration by means of a set of neighborhoods. The neighborhood structures and the perturbation move were similar to the one used in Menéndez et al. [9]. The authors' VNS algorithm was confronted with some simple classical greedy algorithms and a modified Clark & Wright Savings algorithm, demonstrating its superiority (from 7.7% up to 102.2% for the largest problem instances with 250 orders).

3   Problem Formulation

We consider a practical example from a large automotive parts warehouse located in southern Poland. Batches in that warehouse are prepared on the basis of customers' individual attributes, location area of items and due dates required. The batching algorithm for the warehouse is not straightforward, so its analysis and possible improvements are out of scope of the presented paper. Thus we will focus only on batch sequencing and route planning for available pickers during a working shift. Since the composition of each batch is known, the tour length (depending on the underlying routing strategy – in our case the largest gap), the search time and the pick time can be calculated beforehand and stored in matrices. Items to be picked are depicted in an order's line along with theirs demanded quantity. Items for a single line are stored in the same location, but different lines must be usually collected in different locations of the same warehouse area.

The main goal is to develop the picking plan that minimizes the number of pickers engaged in picking process in a given area and simultaneously to optimize the route they have to travel when picking items for orders in the batches (completion lists). The main assumption is that items picking for all orders in each completion list must be finished before deadline (beginning of delivery to a customer) and cannot be picked before the completion list is released. Time here is considered discrete, i.e. the planning horizon is divided into several time periods (with a predefined granularity). The mixed integer programming (MIP) model of the problem can be presented as follows:

Data

I – number of batches,
T – number of discrete time periods,
P – total number of pickers,
d_i – due time for batch i,
l_i – number of lines in batch i (different items' locations),
s_i – the earliest possible time for picking batch i (release time),
e_p – performance of picker p (expressed e.g. as the number of order lines per hour),
a_{pk} – 1, if picker p is available in area k; 0 otherwise,
g_{jk} – 1, if items for batch line j are located in area k; 0 otherwise,
b_k – maximum number of pickers that can work in area k,
P_{tk} – set of pickers available in time window t and area k,
$\delta(j, j')$ – distance between locations of batch line items j and j' that are picked in a planned sequence $j, j' \in O_i$, following routing strategy,
O_i – set of lines (item groups) for batch i.

Decision Variables

x_{ipt} – 1, if picker p is due to pick for order i in time t; 0 otherwise,
r_j – time (discrete period) in which line item j is to be picked.

Objective function

$$w_1 \sum_{p=1}^{P} \max_{i=1..I, t=1..T} x_{ipt} + w_2 \sum_{p=1}^{P} \sum_{t=1}^{T} \sum_{i=1}^{I} \sum_{j, j' \in O_i} x_{ipt} \delta(j, j') \rightarrow min \qquad (1)$$

Constraints

$$\sum_{i=1}^{I} x_{ipt} l_i \leq e_p, p \in P_{tk}, t = 1, \ldots, T \qquad (2)$$

$$\sum_{t=1}^{T} \sum_{p \in P_{tk}} x_{ipt} = 1, i = 1, \ldots, I \tag{3}$$

$$\left(x_{ipt} g_{ik}\right) \geq a_{pk}, i = 1, \ldots, I; p \in P_{tk}; t = 1, \ldots, T; k = 1, \ldots, K \tag{4}$$

$$r_j \leq d_i, i = 1, \ldots, I; j \in O_i \tag{5}$$

$$r_j \geq s_i, i = 1, \ldots, I; j \in O_i \tag{6}$$

$$\sum_{p \in P_{tk}} x_{ipt} g_{ik} \leq b_k, i = 1, \ldots, I; t = 1, \ldots, T; k = 1, \ldots, K \tag{7}$$

Goal function (1) optimizes a weighted sum of the number of pickers engaged in the picking process in a given planning horizon and the total length of the routes each picker has to travel in order to collect all necessary items. Weights w_1 and w_2 can be set by a planner, however for the experiments they have been set to 100 and 1 respectively in order to balance both criteria. Constraints (2) ensure that picker performance is not exceeded. This can be expressed in various ways, depending on the picker's performance measure used in a warehouse, or even combination of them (e.g. items quantity and items weight). In this paper we assume an average orders' lines that are collected by a picker as the weight of the items is comparable in all areas that have been used for benchmark instances. According to constraints (3) only one picker can pick for a single batch i starting in time t. In reality we can have orders that may exceed picking performance of any picker, but in such case the order can be divided into few different batches. The location in which picker p can operate is limited by constraints (4). Constraints (5) say that all items (batch lines) must be picked before their due dates and not earlier than the time in which the batch is released – constraints (6). Finally, constraints (7) limit the number of pickers that can work in the same area in the same time.

4   VNS Algorithm

In [4] a genetic algorithm, solving the problem similar to the one described in the previous section, was introduced. However, in the model under consideration each item was treated separately and no order lines were considered, which occurred to be unrealistic assumption in practical application.

Nevertheless, it was possible to adapt representation structure of a solution (Fig. 1). The solutions are represented by the set of two vectors π and τ, each of the length equal to the number of batches that have been released for picking in a given planning horizon. Vector π represents picker's indices that are engaged in picking batch i and vector τ represents the time (approximated by discrete time periods with a determined precision – in the presented case 30 min) in which the picker picks batch i.

i	1	2	3	4	5	6	7	8	9	10
π_i	1	2	1	2	1	2	2	3	1	3
τ_i	6:30 7:00	7:30 8:00	9:00 9:30	8:00 8:30	6:00 6:30	8:00 8:30	7:00 7:30	8:30 9:00	8:00 8:30	7:30 8:00

Fig. 1. The representation of solution.

Again four different mutations defined in [4] have been adapted for building neighborhood structures in a VNS algorithm. These are as follows:

$N_1(s)$– changes the picking time to another one within the acceptable range, i.e. between the availability time of batch i and its required collection time,
$N_2(s)$– changes picker's index to a different index,
$N_3(s)$– swaps the picking time for the two randomly chosen batches. If the picking time (more precisely: the beginning of the discrete time period) is smaller than release time of the batch, it is changed to the latter time; the same is done if the time (the end of the discrete time period) is greater than the required collection time,
$N_4(s)$– swaps pickers' indexes for the two batches.

The main procedure of the VNS algorithm is shown in Algorithm 1. It follows the basic VNS scheme with the first improvement strategy proposed by Mladenovic and Hansen [11]. In the initial phase, for each batch to be collected a picker's index is drawn in a random way from the available range and the time period is drawn as a period between release time and due time for the batch.

Algorithm 1. VNS algorithm.

```
 1:  for i:=1 to maxinit
 2:    initialize(s)
 3:    s':=local_search(s);
 4:    if f(s') < f(sbest)
 5:      sbest:=s'
 6:  end for
 7:  for i:=1 to maxiter
 8:  for n:=1 to 4
 9:    generate solution s' from neighborhood Nn
10:    s'':=local_search(s')
11:    if f(s'') < f(sbest)
12:      sbest:=s''
13:      n:=1
14:    else
15:      n:=n+1
16:  end for
17: end for
```

Local search procedure is based on the first improvement strategy and it is performed in two steps. At first, the picker with the lowest number of batches that are scheduled to pick is found, then it is replaced with another picker, starting from the one that is involved in picking for the highest number of batches. The procedure continuous for all possible pickers. In the second step, the standard 2-Opt algorithm is used to optimize the routes for all the pickers remaining from the first stage. The local search procedure is summarized in Algorithm 2.

Algorithm 2. Local search for VNS.

1: sort all pickers according to the number of batches planned to be picked by them
2: $i:=1$
3: **repeat**
4:    take picker p_i from the sorted list
5:    $j:=P$;
5:    **repeat**
6:       take picker p'_j from the sorted list
7:       replace p_i with p'_j
8:       **if** performance of p'_j is not violated
9:          $success:=$true
10:      **else**
11:         restore original p_i
12:   **until** $success$ **or** there are no other pickers in the list
13: **until** $success$
14: **for** each picker p_i with batches assigned
15:    **for** all items locations l_{pi} assigned for p_i
16:       2-Opt(l_{pi})
17:    **end for**
18: **end for**

The initial procedure is repeated max_{init} times in order to find a relatively good starting solution. Then for max_{iter} times the procedures of systematic search – using the four defined neighborhoods of the solution – are performed. The sequence of the four neighborhoods and the max_{init} parameter has been fixed to 1000 on the basis of Taguchi's DoE procedure (max_{init} from 100 to 2000 with step 100). The max_{iter} parameter was set to 10000, which allowed to limit the algorithm's execution time for 1, 3 and 5 min, depending on the size of the instances. Limiting the execution time to only few minutes was a crucial requirement due to the fact that the batch (completion) list remains open and new orders for collection may appear during the day and the picking plan has to be quickly reoptimized.

Three sizes of the problem instances have been used as a benchmark for the proposed VNS algorithm. 10 instances in each size have been generated (giving 30 test instances in total), basing on the real data from the warehouse mentioned earlier in Sect. 3. The characteristic of the instance groups is shown are shown in Table 1.

In each case the planning horizon was 8 h (one working shift), time granularity was set to 15 min. It was set in such a way that CPLEX Solver (version 12.8) was able to solve the problem for the smallest instance. For the instances with more batches and more pickers it was impossible for CPLEX to find any feasible solution even after one hour (machine with Intel Xeon 1220 v2 and 16 GB RAM).

Table 1. Characteristics of the three instances groups.

Problem instances	No of batches (completion lists)	No of order lines to be collected	No of pickers available
Small	20	30	5
Medium	50	270	30
Large	80	520	40

For comparison, in addition to CPLEX Solver, we used the genetic algorithm similar to the one presented in [4], which achieved very good results compared to the actual picking plans in the analyzed warehouse (up to 71.2%). The VNS algorithms described in the literature, presented in Sect. 2, could not be used for the comparison. First of all in most cases the presented algorithms were designed exclusively to minimize tardiness regarding picking time, without taking into account the number of pickers. Moreover, they refer primarily to the order batching step, which is not considered in this paper. For example four out of six neighborhood structures in the Scholz et al. VNS algorithm [13] are constructed around moving one or two orders from one batch to the other. The remaining two neighborhoods relay on the swapping batches between pickers (we use similar N_2 and N_4 neighborhoods). Their routing optimization procedure involves either Lin-Kernighan-Helsgaun (LKH) heuristic, 3-Opt or 2-Opt, and the latter is used in our algorithm in order to maintain short execution time of the VNS.

In the experiments computational time for the GA was limited to 1 min, 3 min and 5 min for the small, medium and large instances, respectively, i.e. it was the same as for the VNS algorithm. Additionally the results with the VNS algorithm without local search procedure is presented to show the impact of the LS on the algorithm. Each of the instance was calculated 20 times and average result was taken as the final result for each algorithm. Summary of the results is presented in Table 2. For each approach the average number of pickers engaged and total distance travelled by them are shown.

Table 2. Average results for the compared algorithms.

Instances	VNS		VNS-wLS		GA		CPLEX	
Small								
average	5.0	2482.3	5.0	3090.3	5	2512.2	5.0	2465.7*
std. deviation	0.0	54.8	0.0	73.2	0	88.5	-	-
Medium								
average	22.9	5834.0	29.2	6938.0	23.7	5980.3	N/A	N/A
std. deviation	1.8	127.4	0.5	46.8	1.9	146.1	-	-
Large								
average	33.3	9611.8	40.0	11669.2	34.8	10427.1	N/A	N/A
std. deviation	2.1	138.2	0.0	53.5	2.9	194.4	-	-

As we can observe in the Table 2, the proposed VNS algorithm was better than genetic algorithm in the assumed time limits. For the smallest instances CPLEX Solver gave the best results, however for one instance it was unable to find any feasible solution after 1 h. Results achieved by both VNS and GA were close to the one found by CPLEX Solver. For the medium instances of the problem in most cases VNS was able to plan little less than 23 pickers on average, contrary to the GA which usually planned almost 24 pickers with longer total distance (146 m on average) to be covered by them. A similar advantage of VNS occurred in the case of the largest instances, but the difference between VNS and GA is even larger, especially regarding the distance the pickers have to travel (816 m on average).

We can also see that the local search procedure has a huge impact on the results of the VNS algorithm. Without it the algorithm usually used all available pickers, that additionally have to travel a much longer distance. Its use, however, is very time expensive – it increases the calculation time by over 6 times.

5   Conclusions and the Future Work

Batch sequencing and picker routing problem with time window is one of the most practical problems that is faced in warehouse logistic systems. The paper proposes an efficient VNS approach to minimize the pickers effort.

The contributions of this study are as follows:

- We suggest a new mathematical programing formulation to this problem. A mathematical model was developed to minimize a weighted sum of the pickers engaged in the picking process in a given planning horizon and the total length of the routes the pickers have to travel in order to pick all items. The warehouse studied employs traditional wave picking with multiple pickers.
- We propose a variable neighborhood search able to deal with large instances.
- We conduct numerical experiments to evaluate the performance of our approach.

The computational experiments presented in the paper prove that the proposed VNS based approach can be well applied to the batch sequencing and picker routing problem. The performance of the proposed heuristic depends strongly on the time granularity (discrete time periods) that is used for planning. It is up to the decision maker how small it should be. It is a very important, because smaller time windows lead to the increase of the number of workers involved in the picking process and to the decrease of the average reduction of the routes the pickers have to travel. In our experiments we assumed 15 min, which in practical applications is usually enough.

In the future research better local search algorithm may be considered, allowing for faster (and maybe even better) reduction of the number of pickers that must be engaged in the picking process. Also benefits from using 3-Opt or other heuristics developed for TSP like LKH should be carefully analyzed in relation to the increase of the overall execution time of the VNS algorithm.

Acknowledgement. This study was conducted under a research project funded by a statutory grant of the AGH University of Science and Technology in Krakow for maintaining research potential.

References

1. Albareda-Sambola, M., Alonso-Ayuso, A., Molina, E., de Blas, C.S.: Variable neighborhood search for order batching in a warehouse. Asia-Pac. J. Oper. Res. **26**(5), 655–683 (2009)
2. Ardjmand, E., Shakeri, H., Singh, M., Bajgiran, O.S.: Minimizing order picking makespan with multiple pickers in a wave picking warehouse. Int. J. Prod. Econ. **206**(C), 169–183 (2018)
3. Cergibozan, C., Tasan, A.S.: Order batching operations: an overview of classification, solution techniques. J. Intell. Manuf. **30**(1), 335–349 (2019)
4. Duda, J., Karkula, M.: A hybrid heuristic for order-picking and route planning in warehouses. In: Carpathian Logistics Congress: Logistics, Distribution, Transport & Management CLC 2018, 3 December (2018)
5. Gil-Borrás, S., Pardo, E.G., Alonso-Ayuso, A., Duarte, A.: New VNS variants for the online order batching problem. In: Sifaleras, A., Salhi, S., Brimberg, J. (eds.) ICVNS 2018. LNCS, vol. 11328, pp. 89–100. Springer, Cham (2019). https://doi.org/10.1007/978-3-030-15843-9_8
6. van Gils, T., Ramaekers, K., Caris, A., de Koster, R.B.M.: Designing efficient order picking systems by combining planning problems: State-of-the-art classification and review. Eur. J. Oper. Res. **267**(1), 1–15 (2018)
7. Henn, S.: Order batching and sequencing for the minimization of the total tardiness in picker-to-part warehouses. Flex. Serv. Manuf. J. **27**(1), 86–114 (2015)
8. Henn, S., Koch, S., Wäscher, G.: Order batching in order picking warehouse: a survey of solution approaches. Working paper no. 01/2011, University of Magdeburg (2011)
9. Menéndez, B., Pardo, E.G., Duarte, A., Alonso-Ayuso, A., Molina, E.: General variable neighborhood search applied to the picking process in a warehouse. Electron. Notes Discrete Math. **47**, 77–84 (2015)
10. Menéndez, B., Bustillo, M., Pardo, E.G., Duarte, A.: General variable neighborhood search for the order batching and sequencing problem. Eur. J. Oper. Res. **263**, 82–93 (2017)
11. Mladenović, N., Hansen, P.: Variable neighborhood search. Comput. Oper. Res. **24**(11), 1097–1100 (1997)
12. Petersen, C.G., Schmenner, R.W.: An evaluation of routing and volume-based storage policies in an order picking operation. Decis. Sci. **30**(2), 481–501 (1999)
13. Scholz, A., Schubert, D., Wäscher, G.: Order picking with multiple pickers and due dates – simultaneous solution of order batching, batch assignment and sequencing, and picker routing problems. Eur. J. Oper. Res. **263**(2), 461–478 (2017)
14. Žulj, I., Glock, C.H., Grosse, E.H., Schneider, M.: Picker routing and storage-assignment strategies for precedence-constrained order picking. Comput. Ind. Eng. **123**, 338–347 (2018)

Daily Scheduling and Routing of Home Health Care with Multiple Availability Periods of Patients

Mohammed Bazirha[1(✉)], Abdeslam Kadrani[1], and Rachid Benmansour[1,2]

[1] SI2M Laboratory, INSEA, Rabat, Morocco
{mbazirha,akadrani,r.benmansour}@insea.ac.ma
[2] LAMIH UMR CNRS 8201, Valenciennes, France

Abstract. The home health care routing and scheduling problem (HHCRSP) is an extension of the vehicle routing problem with time windows (VRPTW). It consists of providing services operations at patients' homes in case of aging or disabling disease. In this paper, we address the HHCRSP with multiple availability periods of patients, which are considered as soft/flexible time windows. A mathematical model is proposed to define a daily planning by minimizing the total penalized earliness and tardiness of service operations, and caregivers' total waiting time. Taking into account requested services of patients, qualifications and time windows of caregivers, patients' preferences expressed as multiple availability periods. The model is implemented and tested using CPLEX IBM. To deal with large instances a general variable neighborhood search (GVNS) based heuristic is proposed, implemented and tested using the language C++. Computational results show that the proposed heuristic could find a good solution in a very short computational time.

Keywords: Variable neighborhood search · Home health care · Routing and scheduling · Multiple time windows · Mathematical modeling

1 Introduction and Literature Review

The home health care (HHC) consists of providing care services at patients' home in case of illness, injury or aging in a personal environment [7]. These services include: (i) medical care such as the nursing and the physical therapy, (ii) and non-medical care, including social services and assistances, such as cleaning, preparing food and getting out of bed. Due to the variety of the offered services, people with many different qualifications are employed as caregivers [19]. The HHC will allow patients to remain in their home and receive care and assistance. One of the most important decisions for HHC companies is to assign caregivers to patients by finding for each one an efficient route schedule to reduce operating cost and to improve customer service quality. This problem is well known as the Home Health Care Routing and Scheduling Problem (HHCRSP) is a

© Springer Nature Switzerland AG 2020
R. Benmansour et al. (Eds.): ICVNS 2019, LNCS 12010, pp. 178–193, 2020.
https://doi.org/10.1007/978-3-030-44932-2_13

combination of two NP-hard problems: the nurse rostering problem [6] and the vehicle routing problem with time windows [5], it can be described as follows: a set of patients, scattered in a geographic area, need care services which must be provided by caregivers. They start from the HHC center, travel to provide services for clients, and return to the center with respect to some constraints such as patients' time windows. The objective is to find a set of routes in order to optimize one or more criteria according to the decision maker choice.

Time windows are considered as an important criterion for measuring the patients' satisfaction, so the decision maker should respect as much as possible these time windows' choices by the patients. Two types of time windows are used in the literature: hard/fixed time windows [1,2,14,20], or soft/flexible [12,16,18,24]. In the first one, the decision maker have to schedule the visit within the time window. In the second one, time windows could not be respected, and delays can be accepted with a penalty cost. Redjem et al. [20] defined a heuristic dealing with a simultaneously visits and possibly in a predefined order where patients are supposed to be assigned to caregivers. The objective is to minimize the traveling and waiting times. The heuristic has two steps. The first step is to search the optimal tours by calculating for each caregiver the shortest travel duration. The second step is the introduction of the precedence and the synchronization constraints. Liu et al. [14] have proposed a Branch and Price algorithm (B&P) for the HHCRSP with lunch break requirements, which has to be scheduled on each tour of the care workers. The objective is to minimize the total travel cost and unvisited clients. The B&P algorithm was tested on both real-life data and randomly generated instances modified from the classical Solomon's VRPTW benchmarks. Mankowska et al. [16] have focused their work on temporal dependencies of services, they defined a mathematical model and a heuristic which dealing with double synchronization as well as pairwise temporal precedence between jobs to minimize the total distance traveled by caregivers, the total tardiness of services operations and the maximal tardiness observed over all services. Randomly generated instances were used to test both mathematical model and proposed heuristic. Trautsamwieser and Hirsch [24] have defined a mathematical model and VNS algorithm to optimize the daily planning with respect to mandatory breaks, feasible assignment, hard time windows and working time restrictions. The algorithm was tested on real-life data sets, the randomly generated data sets were used to benchmark the VNS algorithm and to find tuning parameters. Bertels and Fahle [4] have studied the HHCRSP considering both soft and hard times windows, the first included in the second. A hybrid algorithm was used to solve the problem combining linear programming (LP), constraint programming (CP) and simulated annealing (SA) or tabu search (TS). Tricoire et al. [25] and Belhaiza et al. [3] have proposed respectively a heuristic and a hybrid variable neighborhood tabu search heuristic to solve the Vehicle Routing Problem with Multiple Time windows.

Decerle et al. [8] Have proposed a mixed integer-programming model and a memetic algorithm to deal with home health care routing scheduling problem with soft time windows and possible double synchronized services for patients.

Benchmark instances from the literature as well as new instances based on real life data are used to test the proposed methods. Rest and Hirsh [21]: have focused their word on daily scheduling of home health care services considering a multimodal transportation network. A mathematical model and Tabu Search are proposed to deal with the problem, time-dependent travel time is computed by a dynamic programming approach, to optimize travel and waiting times of care staff. Deterministic models and methods are usually less robust. In case of any possible changes in practical situations, the planned scheduling must be redone. Shi et al. [22] proposed a stochastic Programming model, Simulated Annealing, Tabu Search, and Variable Neighborhood Search to solve the problem with stochastic travel and service times. Liu et al. [14] have combined A branch-and-price (B&P) algorithm and a discrete approximation method to solve the Home health care problem where caregiver's travel times and patients' service times are stochastic.

The contribution of this work consists of developing a new mathematical model for the HHC dealing with soft/flexible multiple time windows with a maximum of earliness(E_{max}) and a maximum tardiness(T_{max}) of services operations, where patients could define all periods in which they are available to receive care services. A soft/flexible time windows would increase the chance of finding a feasible schedule as delays are accepted. However, any delays occurred in the case of hard/fixed time windows, the schedule is infeasible and the company must use, maybe hire, more caregivers, which are an additional cost. The soft/flexible time windows proposed with E_{max} and T_{max} is a general case of hard/fixed time windows($E_{max} = 0$, $T_{max} = 0$) and of soft/flexible time windows($E_{max} = \infty$, $T_{max} = \infty$).

Three objectives are considered. The first goal is to minimize the earliness of operations services to ensure patients' availability. The second objective is to minimize the tardiness of operations services to avoid unnecessary waiting times of the patients [16]. The third goal is to minimize caregivers' waiting times, which is considered as an unproductive time [20]. Another distinguishing characteristic of this search is the proposed solving approach, we have developed a metaheuristic, based on General Variable Neighborhood Search, to reduce the computational time needed to solve the model.

2   Problem Statement

Given a set of patients denoted by $N = \{1, 2, 3, ..., n\}$ where n is the number of available patients scattered in a geographic area. Patients need some services denoted by $S = \{1, 2, 3, ..., q\}$, where q is the number of available services provided by the company of HHC. Each patient $i \in N$ has a service duration t_{is}, which it depends on the requested service s. The patient i can specify a multiple times windows $[a_{il}, b_{il}]$, where a_{il} and b_{il} are the earliest and latest service times at his availability period $l \in L = \{1, 2, 3, ..., p\}$ in which he will be available to receive care service. However, only one period will be selected to provide the requested service operation.

Caregivers are denoted by a set $K = \{1, 2, 3, ..., c\}$ where c is the number of available caregivers. They are characterized by a subset of skills S. Each one has a time window $[d_k, e_k]$ where d_k and e_k are the earliest and latest service times. They start and finish their tours at the center of HHC organization, which will be represented by artificial nodes 0 and $n + 1$. The travel time from the patient i to the patient j is denoted by T_{ij}.

Each patient requires one service. Patients' requested services are expressed by a matrix of binary parameters δ_{is}, which is equal to 1 if the patient $i \in N$ requires a service operation $s \in S$, and 0 otherwise. Each service operation s will be assisted only by the caregiver k who have that qualification. Caregivers' skills are expressed by a matrix of binary parameters Δ_{ks}, which is equal to 1 if the caregiver $k \in K$ is qualified to perform a service operation $s \in S$, and 0 otherwise.

The problem is to define a daily planning by minimizing caregivers' waiting time and respecting as much as possible selected patients' time windows, one period for each patient. While assigning qualified caregivers to patients and determining for each one a route such that each caregiver leaves from the HHC center, to serve assigned patients, must to return within their time windows without exceeding the maximum earliness, the maximum tardiness and the maximum waiting time fixed by the decision maker.

The main hypotheses of this problem are:

- The HHC center provides a set of services operations;
- Caregivers start and finish tours at the HHC center;
- Caregivers depart as they are available from the HHC center, i.e. waiting at HHC center is not allowed;
- Each caregiver has a time window and a subset of skills that he could provide;
- Each patient requests a single service and has multiple availability periods;
- Patients' time windows are soft/flexible, it can be violated with a penalty cost;
- A maximum of earliness and tardiness of operations services which not to be exceeded is fixed;
- A maximum waiting time which a caregiver have not to exceed is fixed;
- Processing times of services operations are known and without preemption;
- The travel time between patients are known.

3   Mathematical Formulation

3.1   Parameters

The notation of parameters used in the model is defined as follows:

- α, β, γ: the weights respectively, of total earliness and total tardiness of service operations, and caregivers' total waiting time where $\alpha + \beta + \gamma = 1$;
- N: set of patients;
- $N^0 = N \cup \{0\}$ and $N^{n+1} = N \cup \{n + 1\}$: set of patients including the HHC center which is represented by artificial nodes 0 and $n + 1$;

- n: number of patients;
- K: set of caregivers;
- c: number of caregivers;
- S: set of services (skills);
- q: number of services (skills);
- M: big number;
- L: set of patients' times windows (availability periods);
- p: number of availability periods;
- l: index of patients' times windows;
- $[a_{il}, b_{il}]$: the lth availability period of the patient i;
- $[d_k, e_k]$: caregivers' time windows;
- T_{max}: maximal tardiness of a service operation;
- E_{max}: maximal earliness of a service operation;
- W_{max}: maximal waiting time for each caregiver;
- T_{ij}: travel time from the patient i to the patient j;
- t_{is}: processing time of the service operation s at the patient $i \in N$;
- δ_{is}:equals to 1 if a patient $i \in N$ requires service the operation $s \in S$;
- Δ_{ks}:equals to 1 if the caregiver $k \in K$ is qualified to provide the service operation $s \in S$.

3.2   Decision Variables

The notation of decision variables used in the model is defined as follows:

- x_{ijk}: binary, 1 if the caregiver k visits the patient j after the patient i, 0 otherwise;
- y_{iks}: binary, 1 if the service operation s is provided by the caregiver k to the patient i, 0 otherwise;
- z_{il}: binary, 1 if the lth availability period will be chosen for the patient i, 0 otherwise;
- u_i: earliness of a service operation at the patient i;
- v_i: tardiness of a service operation at the patient i;
- A_{ik}: arrival time of the caregiver k to the patient i;
- S_{ik}: start time of a service operation at the patient i provided by the caregiver k;
- W_{ik}: waiting period of the caregiver k at the patient i;
- W_k: total waiting time of the caregiver k.

3.3   Mathematical Model

The MILP formulation of the problem statement, is an extension of VRPTW [23] adapted and augmented by constraints that are specific to the HHC context, is defined as follows:

$$\min \quad Z = \sum_{i=1}^{n}(\alpha u_i + \beta v_i) + \sum_{k=1}^{c} \gamma W_k$$

$s.t.$

$$\sum_{i=0}^{n}\sum_{k=1}^{c} x_{ijk} = 1, \quad j \in N \tag{1}$$

$$\sum_{j=1}^{n+1}\sum_{k=1}^{c} x_{ijk} = 1, \quad i \in N \tag{2}$$

$$\sum_{i=0}^{n} x_{i(n+1)k} = 1, \quad k \in K \tag{3}$$

$$\sum_{j=1}^{n+1} x_{0jk} = 1, \quad k \in K \tag{4}$$

$$\sum_{i=0}^{n} x_{imk} = \sum_{j=1}^{n+1} x_{mjk}, \quad m \in N, k \in K \tag{5}$$

$$S_{ik} + \sum_{s=1}^{q} t_{is} y_{iks} + T_{ij} \leq S_{jk} + (1 - x_{ijk})M, \quad i \in N^0, j \in N^{n+1}, k \in K \tag{6}$$

$$S_{ik} + \sum_{s=1}^{q} t_{is} y_{iks} + T_{ij} \leq A_{jk} + (1 - x_{ijk})M, \quad i \in N^0, j \in N^{n+1}, k \in K \tag{7}$$

$$S_{ik} + \sum_{s=1}^{q} t_{is} y_{iks} + T_{ij} \geq A_{jk} + (1 - x_{ijk})M, \quad i \in N^0, j \in N^{n+1}, k \in K \tag{8}$$

$$W_{ik} = S_{ik} - A_{ik}, \quad i \in N, k \in K \tag{9}$$

$$W_{ik} \leq \sum_{s=1}^{q} y_{iks}M, \quad i \in N, k \in K \tag{10}$$

$$S_{ik} \leq \sum_{s=1}^{q} y_{iks}M, \quad i \in N, k \in K \tag{11}$$

$$\sum_{j=1}^{n+1} x_{ijk} = \sum_{s=1}^{q} y_{iks}, \quad i \in N, k \in K \tag{12}$$

$$2y_{iks} \leq \delta_{is} + \Delta_{ks}, \quad i \in N, s \in S, k \in K \tag{13}$$

$$S_{0k} = d_k, \quad k \in K \tag{14}$$

$$A_{(n+1)k} \leq e_k, \quad k \in K \tag{15}$$

$$(z_{il} + \sum_{s=1}^{q} y_{iks} - 2)M + a_{il} - u_i \leq S_{ik}, \quad i \in N, l \in L, k \in K \tag{16}$$

$$S_{ik} + \sum_{s=1}^{q} t_{is} y_{iks} \leq b_{il} + v_i + (2 - z_{il} - \sum_{s=1}^{q} y_{iks})M, \quad i \in N, l \in L, k \in K \tag{17}$$

$$\sum_{l=1}^{p} z_{il} = 1, \qquad i \in N \tag{18}$$

$$u_i \le E_{max}, \qquad i \in N \tag{19}$$

$$v_i \le T_{max}, \qquad i \in N \tag{20}$$

$$W_k = \sum_{i=1}^{n} W_{ik}, \qquad k \in K \tag{21}$$

$$W_k \le W_{max}, \qquad k \in K \tag{22}$$

$$x_{iik} = 0, \qquad i \in N, k \in K \tag{23}$$

$$W_{ik} \ge 0, \qquad i \in N, k \in K \tag{24}$$

$$S_{ik} \ge 0, \qquad i \in N, k \in K \tag{25}$$

$$A_{ik} \ge 0, \qquad i \in N, k \in K \tag{26}$$

$$u_i \ge 0, \qquad i \in N \tag{27}$$

$$v_i \ge 0, \qquad i \in N \tag{28}$$

$$x_{ijk} \in \{0,1\}, \qquad i \in N, j \in N, k \in K \tag{29}$$

$$y_{iks} \in \{0,1\}, \qquad i \in N, k \in K, s \in S \tag{30}$$

$$z_{il} \in \{0,1\}, \qquad i \in N, l \in L \tag{31}$$

The objective function is to minimize the total penalized earliness and tardiness of services operations, and caregivers' total waiting time. Constraints (1) and (2) state that each patient will be visited exactly by one caregiver. Constraints (3) and (4) state that each caregiver left the center must to return. Constraints (5) express the flux conservation. Constraints (6) determine the service operations' starting time of the patient j with respect to service operations' completion time of the patient i. These constraints enforce that the starting time of services along the route of a caregiver are strictly increasing. In doing so, they also eliminate sub-tours because a return to an already visited patient would violate the start time of the previous visit [16]. Constraints (7) and (8) define the arrival time of a caregiver k to the patient j. Constraints (9) define the waiting time of the caregiver k at the patient i. Constraints (10) and (11) initialize the waiting time and the starting time to zero if the caregiver k will not be affected to the patient i. Constraints (12) define the variable y_{iks}. Constraints (13) ensure that a qualified caregiver k performs a requested service operation s to patient i. Constraints (14) and (15) enforce the respecting of caregivers' time windows. Constraints (16) and (17) ensure the respecting of patients' time windows. Constraints (18) guarantee that a one period time is selected of the patient's availability periods. Constraints (19) and (20) guarantee not to exceed the maximal earliness and tardiness of a service operation. Constraints (21) define the total waiting time for each caregiver. Constraints (22) ensure not to exceed the maximal waiting time for each caregiver. Constraints (23 to 31) set the domains of the decision variables.

4   Variable Search Neighborhood

Due to the weakness of local search strategies that fall into a local optimum and have no ability to leave it, several metaheuristics, extend the local search strategies, has been proposed to avoid being trapped in a local optimum such as Tabu Search (TS) [11], Simulated Annealing (SA) [13] and Variable search neighborhood (VNS). The VNS was proposed by Mladenovic and Hansen [17], is based on the idea of systematic changes of neighborhoods structure in the search for a better solution. VNS proceeds by a descent method exploring a series of predefined neighborhood to find a local minimum. Each time a starting point is generated randomly using the current neighborhood in the shaking phase to run a local descent method. The new local minimum found is compared to the incumbent, the search jumps to the new solution if and only if is better. Hence, VNS is not a trajectory following method and does not specify forbidden moves. Many versions of VNS are used in the literature such as: (i) Reduced Variable Search Neighborhood (RVNS) is a pure stochastic search method, only shaking phase is applied without improving the generated solution using a local descend method. (ii) Variable neighborhood descent (VND) is a deterministic version of VNS where all neighborhood defined are applied to the initial solution in a predefined order, if a new better local minimum is found the searching is restarted from the first neighborhood. (iii) And the General Variable Search Neighborhood (GVNS) which is a VNS where the local descend method is replaced by the VND.

4.1   Encoding

A solution will be represented by a matrix where the number of columns equals to the number of patients n. Two lines are used, the first one will contain patients and services operations requested (are included in parenthesis), and the second will contain assigned caregivers. Example: we assume that we have 6 patients and 2 caregivers skilled to provide 3 types of services operations. A solution will be encoded as follow (see Table 1).

The caregiver 1 will visit the patient 1 to provide the service operation 3, the patient 2 to provide the service operation 1 and the patient 5 to provide the service operation 2.

Table 1. Example of solution encoding

Patients	1 (3)	3 (1)	4 (3)	2 (1)	6 (2)	5 (2)
Caregivers	1	2	2	1	2	1

4.2   Decoding

Algorithm 1: Caregivers' arriving early to patients

---

1 **if** $(\alpha \leq \gamma)$ **then**

2 | **if** $(ET_{il} \leq E_{max})$ **then**

3 | | set $u_i \longleftarrow ET_{il}$;

4 | **else if** $(W_k + ET_{il} - E_{max} \leq W_{max})$ **then**

5 | | set $u_i \longleftarrow E_{max}$ and $W_k \longleftarrow W_k + ET_{il} - E_{max}$;

6 | **else**

7 | | the availability period l is infeasible

8 | **end**

9 **else**

10 | **if** $(ET_{il} + W_k \leq W_{max})$ **then**

11 | | set $W_k \longleftarrow W_k + ET_{il}$;

12 | **else if** $(W_k + ET_{il} - W_{max} \leq E_{max})$ **then**

13 | | set $u_i \longleftarrow W_k + ET_{il} - W_{max}$ and $W_k \longleftarrow W_{max}$;

14 | **else**

15 | | the availability period l is infeasible

16 | **end**

17 **end**

Given a solution encoded as proposed above. For each sub set of patients assigned to a caregiver, the starting, the arrival and the waiting times will be calculated iteratively in the same order as they appear at the matrix. For each patient i, the period $l \in L$, constraints (1) to (31) are taken into account, that minimize the waiting time, the earliness and the tardiness of the service operation is selected (see Eq. 32).

$$\arg\min_{l \in L}\{\alpha u_i + \beta v_i + \gamma W_{ik} \mid s.t.\ Constraints\ (1)\ to\ (31)\} \qquad (32)$$

For each period $l \in L$, three possible cases of caregivers' arrival times are to distinguish:

1. The caregiver arrives to a patient and finish providing requested service operation within the availability period:

$$u_i = 0, \quad v_i = 0 \quad and \quad W_{ik} = 0$$

2. The caregiver arrives to a patient within the availability period and finish providing requested service operation with a tardiness time:

$$u_i = 0, \quad v_i = A_{ik} + t_{is} - b_{il} \quad and \quad W_{ik} = 0$$

3. The caregiver arrives to a patient before the availability period . In this case, many possibilities arise to calculate the waiting time and the earliness of the service operation. This problem could be formulated as a MIP problem to determine the optimal combination. However, the Algorithm 1 is used. The early time of the caregiver k at patient i for the availability period l (ET_{il}) is defined by the following formula: $ET_{il} = a_{il} - A_{ik}$.

4.3 Neighborhoods

The neighborhood of a solution is defined as a transformation function applied to this solution to get a set of solutions where one can move some amount in any direction away from that solution without leaving the set. Four neighborhoods structure are proposed, two neighborhoods are used to intensify patients' assignment to caregivers (switch and inter-swap) and the two others are used to intensify the order visiting (intra-shift and intra-swap).

1. Switch (i.e. Patients' reassignment to caregivers): the neighborhood of a solution is defined as a reassignment of another caregiver k to a patient i. The size of possible neighborhoods will be less than $n(c - 1)$ depending on caregivers' qualifications. The equality could be hold if all caregivers are skilled to provide all services operations (see Fig. 1);

Original solution :	Patients	**1 (3)**	3 (1)	4 (3)	2 (1)	6 (2)	5 (2)
	Caregivers	**1**	2	2	1	2	1
Neighbor solution :	Patients	**1 (3)**	3 (1)	4 (3)	2 (1)	6 (2)	5 (2)
	Caregivers	**2**	2	2	1	2	1

Fig. 1. Example of switch neighborhood moves

2. Inter-swap: this neighborhood aims to change patients' assignment to caregivers. Given two patients, caregivers' assignment are swapped. The size of possible neighborhoods equals to $\frac{(n-1) \times n}{2}$ (see Fig. 2);

Original solution :	Patients	**1 (3)**	3 (1)	4 (3)	2 (1)	**6 (2)**	5 (2)
	Caregivers	1	2	2	1	2	1
Neighbor solution :	Patients	**1 (3)**	3 (1)	4 (3)	2 (1)	**6 (2)**	5 (2)
	Caregivers	2	2	2	1	**1**	1

Fig. 2. Example of inter-swap neighborhood moves

3. Intra-shift: given a visiting order, the neighborhood is defined as shifting of a patient to another position. The size of possible neighborhoods equals to $(n - 1) \times n$ (see Fig. 3);

Original solution :	Patients	**1 (3)**	3 (1)	4 (3)	2 (1)	6 (2)	5 (2)
	Caregivers	1	2	2	1	2	1
Neighbor solution :	Patients	3 (1)	4 (3)	2 (1)	**1 (3)**	6 (2)	5 (2)
	Caregivers	2	2	1	**1**	2	1

Fig. 3. Example of intra-shift neighborhood moves

4. Intra-swap: given a visiting order, the neighborhood is defined as two patients' position exchanging. The size of possible neighborhoods equals to $\frac{(n-1) \times n}{2}$ (see Fig. 4).

Original solution :	Patients	1 (3)	3 (1)	4 (3)	2 (1)	6 (2)	5 (2)
	Caregivers	1	2	2	1	2	1
Neighbor solution :	Patients	6 (2)	3 (1)	4 (3)	2 (1)	1 (3)	5 (2)
	Caregivers	2	2	2	1	1	1

Fig. 4. Example of intra-swap neighborhood moves

4.4 Shaking

The shaking phase is the heart of the algorithm which is used to avoid being trapped in a local optimum. It is defined as one or a series of moves applied to a solution to jump from a local optimum. The four proposed neighborhoods are used for the shaking phase ($k_{max} = 4$) as operators which will be applied m times, each time the move is generated randomly. A series of tests was executed to find the best value of m which is fixed to $m = 5$.

4.5 Local Search

Two types of algorithm could be used in local search: first improvement and best improvement, the first one consists of staring over the search when the first neighbor that improve the initial solution is found and the second consists of starting over the search when all neighbors are tested and the best one is selected. Four local search methods will be used where each one is matched to a neighborhood of the four proposed, the best improvement algorithm will be adopted to these local search methods.

4.6 Initial Solution

The initial solution is generated randomly as follow:

1. Patients are sorted by increasing end of their time windows for single period. for multiple periods the visiting order is generated randomly (see Table 2)

Table 2. Example of visiting order

Patients	1 (3)	3 (1)	4 (3)	2 (1)	6 (2)	5 (2)
Caregivers						

2. For each patient, assign a qualified caregiver selected randomly(see Table 3);

Table 3. Example of caregivers' assignment to patients

Patients	1 (3)	3 (1)	4 (3)	2 (1)	6 (2)	5 (2)
Caregivers	1	2	2	1	2	1

3. Calculate caregivers' waiting times, the earliness and tardiness of services operations using the decoding method proposed above.
4. If the solution is infeasible repeat steps 1, 2 and 3. Otherwise, go to the step 5;
5. calculate the objective function value.

4.7   GVNS Algorithm

The stopping condition is fixed as a number of no improvement in the best solution found for X iterations which equals to the number of patients n times 10. Therefore the counter is initialized to zero if a new better solution is found. The VND is applied to each solution x' generated by the shaking phase. each time the solution x (resp. x') is improved the k (resp. l) is initialed to 1.

Algorithm 2: GVNS algorithm

---

1 **Initialization** : ;
2   - set $K_{max} = 4$ and $l_{max} = 4$;
3   - **generate** an initial solution x ;
4 **while** *(the stopping condition is not reached)* **do**
5 | **for** $k \leftarrow 1$ **to** k_{max} **do**
6 | | **generate** at random x' in the kth Neighborhood of x, the kth Neighborhood move is applied 5 times to x ;
7 | | **for** $l \leftarrow 1$ **to** l_{max} **do**
8 | | | **find** the best neighbor x" of x in $N_l(x')$;
9 | | | **if** $f(x'') < f(x')$ **then**
10 | | | | set $x' \longleftarrow x''$ and $l \longleftarrow 1$;
11 | | | **else**
12 | | | | set $l \longleftarrow l + 1$;
13 | | | **end**
14 | | **end**
15 | | **if** $f(x') < f(x)$ **then**
16 | | | set $x \longleftarrow x'$ and $k \longleftarrow 1$;
17 | | **else**
18 | | | set $k \longleftarrow k + 1$;
19 | | **end**
20 | **end**
21 **end**

---

5   Numerical Experiments

The experiments run on the computer with Intel i7–7600U 2.80-GHz CPU and 16 GB of RAM under windows 10. The MIP model is implemented and tested using CPLEX IBM version 12.8. The metaheuristic based on GVNS is coded and tested using the language C++.

5.1   Test Instances

Tests instances have been generated randomly using the benchmark instances from Mankowska et al. [16]. Patients and the HHC office are placed at random locations in the area of 100×100 distance units. Travel times T_{ij} are equal to the Euclidean distance between patients truncated to integer. Processing times of services operations t_{is} are randomly chosen from the interval $[10, 20]$. Six types of services $S = \{1, ..., 6\}$ are considered. Caregivers are grouped into two subsets with different skills. Each caregiver of the first group is qualified for providing at most three services, which are randomly selected from the subset $\{1, 2, 3\}$ of S. Accordingly, each caregiver of the second group is qualified for providing at most three services from subset $\{4, 5, 6\}$. Each patient requires a single service, which is randomly drawn from $S = \{1, ..., 6\}$. The time windows are of length 120 min (2 h) and are randomly placed within a daily planning period of 10 h. Regarding instances which contains two availability periods, the first period will be placed in the first 5 h, and the second period will be placed in the interval $[5, 10]$. No special preference for the three sub-goals and, therefore, weights are set to $\alpha = \beta = \gamma = 1/3$. The maximum earliness and the maximum tardiness of service operations are set respectively to 0 min and 15 min, the maximum waiting time is set to 90 min. 8 instances are generated, each one is used with 1-period and with 2-periods availability of patients. The instance Int1_1 refers to the instance 1 with one availability period and the instance Int1_2 refers to the instance 1 with two availability periods. A series of tests was executed to find the best tuning parameters for the proposed metaheuristic.

5.2   Computational Results

Instances are generated as described above and solved. Instances with two availability periods are generated feasible in the first time. However, instances with single time windows need many regenerations to get a feasible solution. Table 1 summarizes the results of CPLEX and GVNS according to the sizes and patients' availability periods of test instances. 'LB' is lower bound of the model given by CPLEX IBM. 'Z' is the objective function value, 'GAP' is calculated as 100% × ((average − lower bound)/average) and 'CPU' is computing time elapsed of solved instances. For the GVNS algorithm, each instance is running 10 times and the best, the worst and the average solution are considered. 'CPU' computing time correspond to sum of time elapsed to solve each instance 10 times. All instances are solved, using CPLEX IBM, to optimality except instances Int7_2 and Int8_2. Instances Int6_2, Inst7_2 and Int8_2 are hard to solve compared to others instances. The instance Int6_2 required 47530 s to be solved, the instance Int7_2 and Int8_2 was executed for 18 h and CPLEX could not find respectively the optimal solution and a feasible solution, which shows the limit of the exact method used considering the problem is NP-hard. The proposed metaheuristic (GVNS), was able to find a good solution in a very short computational time, solution in bold are proven optimal (see Table 4). The proposed GVNS could find optimal solutions at least one time of the 10 executions for 11 instances.

Table 4. Numerical results of tested instances

| Instances | N | K | L | CPLEX | | | GVNS | | | | |
				LB	Z	CPU	Best	Worst	Average	Gap	CPU
Int1_1	7	2	1	11	**11**	1.58	**11**	**11**	**11**	0%	<1
Int2_1	10	3	1	31.33	**31.33**	1.90	**31.33**	31.36	31.53	1%	<1
Int3_1	14	4	1	24	**24**	2.52	**24**	35.33	27.76	13%	1.79
Int4_1	20	5	1	7	**7**	3.02	**7**	9.33	7.43	6%	8.82
Int5_1	25	6	1	45	**45**	5.16	59.86	84	65.1	30%	25
Int6_1	30	6	1	34.6	**34.6**	12.13	35.3	44.6	39.13	17%	25
Int7_1	40	8	1	54	**54**	255	72	100	78	30%	111
Int8_1	50	10	1	7.66	**7.66**	381	11	18.66	13.73	44%	210
Int1_2	7	2	2	11.33	**11.33**	1.73	**11.33**	**11.33**	**11.33**	0%	<1
Int2_2	10	3	2	5.33	**5.33**	1.91	**5.33**	6.33	5.43	2%	<1
Int3_2	14	4	2	2.66	**2.66**	2.60	**2.66**	4	3.16	16%	1.74
Int4_2	20	5	2	0	**0**	16.31	**0**	9	2.7	100%	9.77
Int5_2	25	6	2	0	**0**	61	**0**	6.33	4.366	100%	32
Int6_2	30	6	2	35.3	**35.3**	47530	**35.3**	44.66	41.96	15%	52.64
Int7_2	40	8	2	0	2.33	64800	5.66	13.66	7.86	100%	248
Int8_2	50	10	2	0	-	64800	**0**	1.33	0.23	100%	561

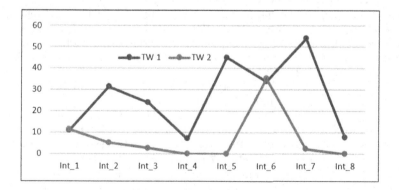

Fig. 5. Objective function values according to the number of time windows (TW)

Due to the LB equals zero, gap of instances Int4_2 and Int6_2 is 100%. Because when LB approaches zero, the gap tends to 100%.

The objective function values of instances with two availability periods are less than those with single period, except the first and the sixth instances, their values are very close (see Fig. 5). Therefore, patients will be more satisfied seeing that the earliness and tardiness of service operations are minimized. On the other hand, the waiting time is also minimized which will increase the productive time of caregivers since the waiting time is considered as unproductive time.

6 Conclusion

In this paper we have addressed the home health care routing and scheduling problem with soft/flexible multiple time windows for patients. The objective is to define a daily planning in which each caregiver would visit assigned patients and return to the HHC center. To do that, a mathematical model is proposed to minimize the total penalized earliness and tardiness of services operations, and the caregivers' total waiting time. The model is tested on randomly generated data. Since exact methods did not provide a good solution in the reasonable computational time, a powerful metaheuristic, based on general variable neighborhood search (GVNS), is proposed to deal with large instances, which could find a good solution in a very short computational time. The soft/flexible time windows would increase the chance of finding a feasible solution since delays are accepted with a penalty cost. However, patients will be less satisfied because their availability periods are not respected. The future work will be focused on hard/fixed multiple time windows for patients and multiple synchronized services.

References

1. Akjiratikarl, C., Yenradee, P., Drake, P.R.: Pso-based algorithm for home care worker scheduling in the UK. Comput. Ind. Eng. **53**(4), 559–583 (2007)
2. Bard, J.F., Shao, Y., Jarrah, A.I.: A sequential grasp for the therapist routing and scheduling problem. J. Sched. **17**(2), 109–133 (2014)
3. Belhaiza, S., Hansen, P., Laporte, G.: A hybrid variable neighborhood tabu search heuristic for the vehicle routing problem with multiple time windows. Comput. Oper. Res. **52**, 269–281 (2014)
4. Bertels, S., Fahle, T.: A hybrid setup for a hybrid scenario: combining heuristics for the home health care problem. Comput. Oper. Res. **33**(10), 2866–2890 (2006)
5. Bräysy, O., Gendreau, M.: Vehicle routing problem with time windows, Part I: route construction and local search algorithms. Transp. Sci. **39**(1), 104–118 (2005)
6. Burke, E.K., De Causmaecker, P., Berghe, G.V., Van Landeghem, H.: The state of the art of nurse rostering. J. Sched. **7**(6), 441–499 (2004)
7. Cissé, M., Yalçındağ, S., Kergosien, Y., Şahin, E., Lenté, C., Matta, A.: Or problems related to home health care: a review of relevant routing and scheduling problems. Oper. Res. Health Care **13**, 1–22 (2017)

8. Decerle, J., Grunder, O., El Hassani, A.H., Barakat, O.: A memetic algorithm for a home health care routing and scheduling problem. Oper. Res. Health Care **16**, 59–71 (2018)
9. Di Mascolo, M., Espinouse, M.L., El Hajri, Z.: Planning in home health care structures: a literature review. IFAC-PapersOnLine **50**(1), 4654–4659 (2017)
10. Fikar, C., Hirsch, P.: Home health care routing and scheduling: a review. Comput. Oper. Res. **77**, 86–95 (2017)
11. Glover, F.: Future paths for integer programming and links to artificial intelligence. Comput. Oper. Res. **13**(5), 533–549 (1986)
12. Hiermann, G., Prandtstetter, M., Rendl, A., Puchinger, J., Raidl, G.R.: Metaheuristics for solving a multimodal home-healthcare scheduling problem. CEJOR **23**(1), 89–113 (2013). https://doi.org/10.1007/s10100-013-0305-8
13. Kirkpatrick, S., Gelatt, C.D., Vecchi, M.P.: Optimization by simulated annealing. Science **220**(4598), 671–680 (1983)
14. Liu, R., Yuan, B., Jiang, Z.: Mathematical model and exact algorithm for the home care worker scheduling and routing problem with lunch break requirements. Int. J. Prod. Res. **55**(2), 558–575 (2017)
15. Liu, R., Yuan, B., Jiang, Z.: A branch-and-price algorithm for the home-caregiver scheduling and routing problem with stochastic travel and service times. Flexible Serv. Manuf. J. **31**(4), 989–1011 (2018). https://doi.org/10.1007/s10696-018-9328-8
16. Mankowska, D.S., Meisel, F., Bierwirth, C.: The home health care routing and scheduling problem with interdependent services. Health Care Manage. Sci. **17**(1), 15–30 (2013). https://doi.org/10.1007/s10729-013-9243-1
17. Mladenović, N., Hansen, P.: Variable neighborhood search. Comput. Oper. Res. **24**(11), 1097–1100 (1997)
18. Nickel, S., Schröder, M., Steeg, J.: Mid-term and short-term planning support for home health care services. Eur. J. Oper. Res. **219**(3), 574–587 (2012)
19. Rasmussen, M.S., Justesen, T., Dohn, A., Larsen, J.: The home care crew scheduling problem: preference-based visit clustering and temporal dependencies. Eur. J. Oper. Res. **219**(3), 598–610 (2012)
20. Redjem, R., Marcon, E.: Operations management in the home care services: a heuristic for the caregivers' routing problem. Flex. Serv. Manuf. J. **28**(1–2), 280–303 (2016)
21. Rest, K.-D., Hirsch, P.: Daily scheduling of home health care services using time-dependent public transport. Flex. Serv. Manuf. J. **28**(3), 495–525 (2015). https://doi.org/10.1007/s10696-015-9227-1
22. Shi, Y., Boudouh, T., Grunder, O.: A robust optimization for a home health care routing and scheduling problem with consideration of uncertain travel and service times. Transp. Res. Part E: Logist. Transp. Rev. **128**, 52–95 (2019)
23. Solomon, M.M., Desrosiers, J.: Survey paper–time window constrained routing and scheduling problems. Transp. Sci. **22**(1), 1–13 (1988). https://doi.org/10.1287/trsc.22.1.1
24. Trautsamwieser, A., Hirsch, P.: Optimization of daily scheduling for home health care services. J. Appl. Oper. Res. **3**(3), 124–136 (2011)
25. Tricoire, F., Romauch, M., Doerner, K.F., Hartl, R.F.: Heuristics for the multi-period orienteering problem with multiple time windows. Comput. Oper. Res. **37**(2), 351–367 (2010)

Author Index

Printed in the United States
By Bookmasters